the science of...
ALiENs

Jack Challoner

August 14, 2007

To my husband, Randy, who is...
out of this world!

Love, Greg & Missy

Georgie Porgie

the science of... ALiENs

Jack Challoner

PRESTEL
MUNICH • BERLIN • LONDON • NEW YORK

This book has been published in conjunction with the exhibition The Science of Aliens,
held at The Science Museum, London, starting on 14 October 2005.

© for the text by Science Museum Solutions
© for design and layout by Prestel Verlag, Munich · Berlin · London · New York 2005
© for illustrations see Picture Credits, page 126
'*Alien Imagination*' chapter produced in association with Channel 4's *Alien Worlds*, a Big Wave Production.
All *Alien Worlds* text and images © Big Wave 2005.

Cover: © Big Wave 2005; background Jerry Schad/Science Photo Library.
Back cover: Alien autopsy exhibit at UFO Museum, Roswell, recreating events subsequent to UFO sightings in the USA in 1947.
Peter Menzel/Science Photo Library.
Front flap: top row MEPL; 2nd row, left to right: Eye of Science/Science Photo Library; Don Perrine/naturepl.com;
Phil Hurst/NHMPL; 3rd row: all © Big Wave Productions 2005; bottom row, left to right: Steve Alexander
www.temporarytemples.co.uk; NASA/HQ/GRIN; NASA/SSPL.

pp. 8–9 The surface of Aurelia closest to its red dwarf sun. Big Wave.
p. 124 Betty Hill, abducted aboard an alien UFO, is shown a 'star map' by her abductors which is later shown to
be accurate, 19 September 1961. Detail of an original painting by Michael Buhler. Mary Evans Picture Library.

Prestel Verlag
Königinstrasse 9,
D-80539 Munich
Tel. +49 (89) 38 17 09-0
Fax +49 (89) 38 17 09-35
www.prestel.de

Prestel Publishing Ltd.
4, Bloomsbury Place,
London WC1A 2QA
Tel. +44 (020) 7323-5004
Fax +44 (020) 7636-8004

Prestel Publishing
900 Broadway, Suite 603,
New York, N.Y. 10003
Tel. +1 (212) 995-2720
Fax +1 (212) 995-2733
www.prestel.com

The Library of Congress Control Number: 2005904404

The Deutsche Bibliothek holds a record of this publication in the Deutsche Nationalbibliographie; detailed bibliographical
data can be found under: http://dnb.dde.de

Prestel books are available worldwide. Please contact your nearest bookseller or one of the above addresses for information
concerning your local distributor.

Editorial direction: Philippa Hurd
Design, typesetting and layout: Sugarfree Design Limited, London. sugarfreedesign.co.uk
Picture research: Kate Pink
Origination: Wyndeham Kestrel, Essex, UK
Printing and binding: Print Consult, Munich

Printed in Slovakia on acid-free paper.
ISBN 3-7913-3485-9

Foreword
by Professor Martin Rees
Astronomer Royal

Does alien life exist? This is surely one of the most fascinating questions in the whole of science. I am hopeful that we shall learn the answer by the end of this century.

In earlier centuries, many believed that the Moon and Mars were inhabited. The science fiction of Jules Verne and H. G. Wells popularised the idea of alien life. Today we are less optimistic about Mars than our forbears were 100 years ago, as we know that there is certainly nothing like the Martians of popular fiction on that planet. An armada of space probes is being launched to analyse the surface of Mars, to fly over it, and (in later missions) to return samples to Earth. Life could also exist in the ice-covered oceans of Jupiter's frozen moon, Europa, and there are plans to land a submersible probe that could explore beneath the ice. Detection of even the most primitive life forms would be a great discovery, as it would offer clues to the mystery of how life began.

Not even the optimists expect to find 'advanced' life elsewhere in our Solar System. But our Sun is just one star among billions, and in the vastness of space far beyond our own Solar System we can rule out nothing. Astronomers have discovered that other stars have their own retinue of planets around them, just as the Earth, Mars and Jupiter circle the Sun. Could some of these planets, orbiting other stars, harbour life forms far more interesting and exotic than anything we might find on Mars? Could they even be inhabited by beings that we would recognise as intelligent?

Claims that advanced life is widespread must confront the famous question first posed by the great Italian physicist Enrico Fermi: if intelligent aliens are common, should they not have visited us already? Why are they, or their artefacts, not staring us in the face? Should we not have seen so many UFOs that there is absolutely no doubt about their existence? This argument gains further weight when we realise that some stars are billions of years older than our Sun: if life were common, its emergence should have had a head start on the planets that orbit these ancient stars.

But the fact that we have not been visited does not imply that aliens do not exist. It would be far harder to traverse the mind-boggling distances of interstellar space than to transmit a signal.

That is perhaps how aliens would reveal themselves first. Searches for extraterrestrial intelligence (SETI) using large radio telescopes have concentrated on 'listening' for radio transmissions that could be artificial in origin. This option is familiar from fictional depictions, such as Carl Sagan's novel *Contact*. Short stretches of data from the SETI searches have been downloaded by millions of people to use as screen savers on their home computers, each hoping to be the first to detect ET.

If we found such a signal, could we build up communication? Intelligent aliens would probably be hundreds of light years away, or more. Can we communicate with beings whose messages may take hundreds, thousands, even millions of years to reach us? The outcome is uncertain, but I am enthusiastic about these searches, because of the import of any manifestly artificial signal. Even if we could not make much sense of it, we would have learnt that 'intelligence' was not unique and had emerged elsewhere. Our cosmos would seem far more interesting; we would look at a distant star with renewed interest if we knew it was another Sun, shining on a world as intricate and complex as our own.

Even if these searches fail, that does not mean that we are alone. The brains and senses of the aliens may be so different from ours that we could not recognise any patterns in their signals; or they may not be transmitting at all. The only type of intelligence we could detect would be one that led to a technology that we could recognise, and that could be a minor and atypical fraction. Some 'brains' may have a quite different perception of reality. Super-intelligent dolphins could be enjoying a contemplative life on some water-covered planet without us even knowing. Still other 'brains' could actually be assemblages of 'social insects'. If evolution on another planet in any way resembled the 'artificial intelligence' scenarios conjectured for the 21st century here on Earth, the most likely and durable form of 'advanced life' may be machines whose creators may long ago have been usurped or become extinct. There may be a lot more out there than we could ever detect, or even imagine. Absence of evidence is not evidence of absence.

Fictional aliens are sometimes cute, sometimes scary, and usually depicted as mammalian bipeds. But the reality, as this book shows, could be far more varied and exotic. There is an enormous variety of life on Earth, from slime mould to monkeys (and, of course, humans). Far greater variety could exist elsewhere in the Galaxy: huge bulbous creatures floating in the dense atmospheres of Jupiter-like planets; aliens the size of insects on a planet where the force of gravity is very strong; or aliens that float freely in space. The great astronomer Fred Hoyle wrote a classic science-fiction novel called *The Black Cloud,* in which a cosmic cloud permeated by swirling electric currents behaves like a super-intelligent brain.

We know too little about how life began, and how it evolves, to be able to say whether alien intelligence is likely or not. Indeed we do not know what would have happened if, as it were, the tape were re-run on Earth. If asteroid impacts and volcanic eruptions had not happened the way they did, would the Earth have ended up harbouring intelligent reptiles, or just insects, or would there have been a convergent evolution towards something humanoid? The cosmos could already be teeming with life: if so, nothing that happens on Earth would make much difference to life's long-range cosmic future. On the other hand, the emergence of intelligence may require such an improbable chain of events that it is unique to our Earth.

But even if aliens do not exist at the moment, they may exist in the distant future. It has taken nearly 4 billion years for humans to evolve from the first life on Earth. But our Sun has burnt less than half its nuclear fuel supply, and it will be another 6 billion years before it flares up and dies. That allows time for descendants of the human species to evolve, both here on Earth and maybe far beyond, into creatures as different from us as we are from protozoa. The aliens of our imagination could exist far away in space, or they could resemble our own post-human descendants.

The idea of life existing elsewhere in the Universe has been put forward by certain philosophers as far back as the time of Ancient Egypt. Serious philosophical debate about 'the plurality of worlds' – the idea that Earth is not the only inhabited planet – stretches back at least to Ancient Greece.

In addition to the stars, astronomers in Antiquity knew, of course, of the Moon and the Sun, and the planets as far out as Saturn, though they did not know that planets were anything more than lights wandering slowly across the night sky. Some did know, however, that the Moon is spherical. All of the early literature about extraterrestrials referred to creatures living on the Moon, often called Selenites. The earliest known story featuring aliens was written in the 2nd century by the Greek satirist Lucian of Samosata. In *Vera Historia* ['True Story'], a storm sweeps up Lucian's ship and carries it to the Moon, where he finds its inhabitants to be a warlike but cultured 'people'.

The ideas of the Greek philosopher Aristotle stifled debate about the possibility of extraterrestrial life. Aristotle argued

▲ Nicolaus Copernicus (left), whose heliocentric model of the Universe was given support by the observations of Galileo Galilei (middle). Johannes Kepler (right) showed that planetary orbits around the Sun are ellipses, not perfect circles.

Figure: French print from 1791, whose title translates as 'Fontenelle Pondering the Plurality of Worlds'. Fontenelle's extremely popular book *Discussions on the Plurality of Worlds* opened people's minds to the possibility of extraterrestrial life.

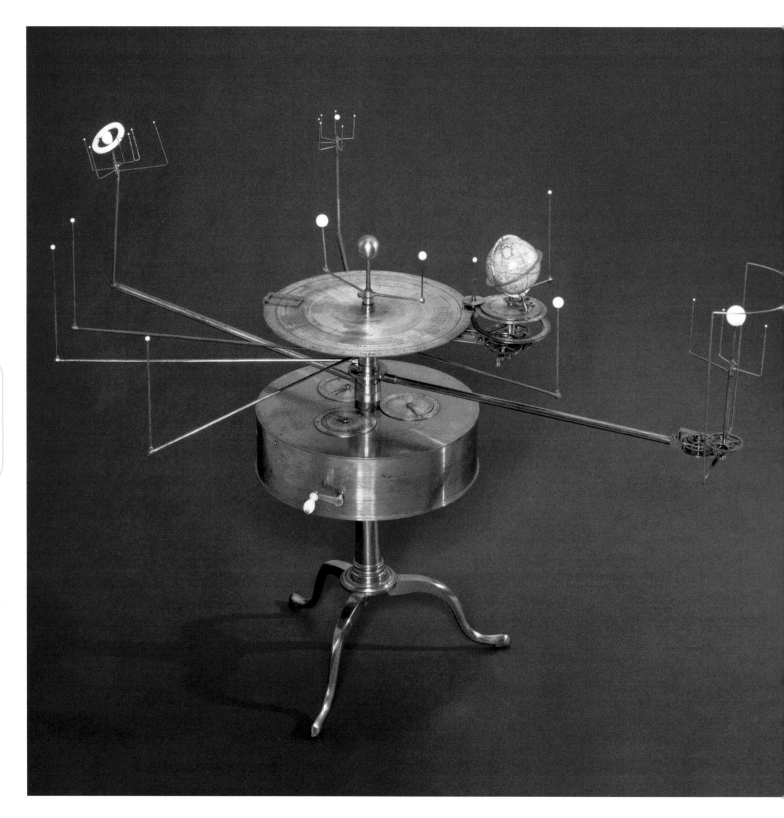

▲ An orrery – a mechanical model of the Solar System – made between 1813 and 1822. The sizes and distances of the planets are not to scale, but a small crank handle drives the planets around the Sun at the centre.

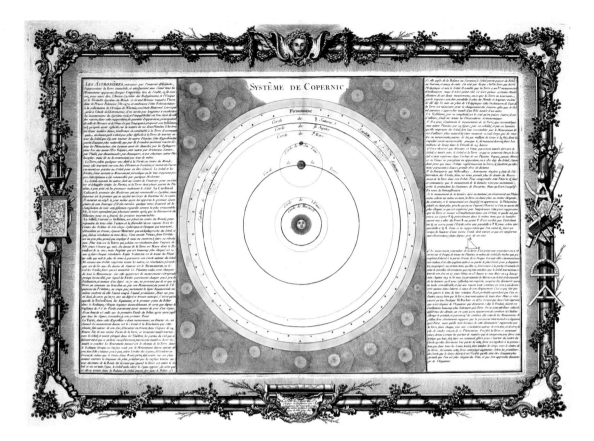

strongly against the plurality of worlds, maintaining that God must have created human beings in a privileged position, at the centre of the Universe, with the Earth as the only place where people could live. While there was some resistance to this idea, Aristotle's views largely prevailed until the middle of the 16th century, when Polish astronomer Nicolaus Copernicus provided convincing evidence that Earth and the other planets orbit the Sun – effectively shifting the perceived centre of the Universe to the Sun and not the Earth. When astronomers began turning telescopes to the sky in the 17th century, they discovered surface details on the Moon, and moons orbiting Jupiter and Saturn. As a result of the new heliocentric view of the Universe, the idea of extraterrestrial life came to the forefront again: if Earth is

just a planet, then the other planets might be similar to Earth, and might harbour life.

With this new interest in space, stories about space travel and extraterrestrial life became more common during the 17th and 18th centuries, though they were still limited almost exclusively to the Moon. In 1634 German astronomer Johannes Kepler wrote of an imaginary voyage to the Moon in *Somnium* ['The Dream']; four years later, Francis Godwin, Bishop of Hereford, wrote *The Man in the Moone*, about a utopian lunar society. In 1657, Cyrano de Bergerac's *A Comical History of the States and Empires of the Moon*, was published posthumously, followed by *A Comical History of the States and Empires of the Sun* in 1662. In 1686, Bernard le Bovier de Fontenelle wrote one of the best-

Eighteenth-century French print illustrating Nicolaus Copernicus' heliocentric view of the Universe. First published in 1543, Copernicus' view made it seem likely that other solar systems exist, stimulating interest in the possibility of extraterrestrial life.

The Science of Aliens

selling and most influential books of its time, *Entretiens sur la pluralité des mondes* ['Discussions on the Plurality of Worlds'], which argued strongly in favour of other worlds. In the preface he writes: 'It seems to me that nothing could be of greater interest than to know how this world we inhabit is made, and if there are other worlds which are similar to it which are inhabited, too.' In the 18th century, French author and philosopher Voltaire wrote *Micromégas* (1752), the first story to include a visit to Earth by aliens. There are two aliens in Voltaire's story – one from Saturn and one from a planet orbiting the star Sirius – who have come to mock human beings for their stupidity. This theme has recurred in many science fiction stories since.

The remarkable pace of innovation in the 19th century gave rise to the first true science fiction: stories that made use of real scientific theories to create imaginative tales. That century saw the creation of industrial steel and synthetic rubber; the rise of steam ships and the railway; the invention of photography, the telegraph and then the telephone; and the first sound recordings. Discoveries in geology led to a new realisation: that the Earth – and therefore the Universe – is incredibly old. Experiments with electricity and magnetism led to a unified understanding of energy and matter. All of these things fed into the developing genre of science fiction. When it came to writing about alien life, it was the discoveries of 19th-century astronomers that inspired science-fiction writers to imagine extraterrestrial life.

With new, large, powerful telescopes, 19th-century astronomers could peer deeper into space. Using a technique known as spectroscopy, they began to find out the chemical compositions of stars and planets. By attaching photographic plates to their telescopes, they could gather more light, and produce exciting images of the planets. And for the first time, astronomers could work out the distances to certain stars. The Universe, they realised, is extremely large, and the Sun is just an ordinary star, like countless others.

During much of the 19th century, the Moon was still the favoured location of those who believed in the possibility of alien life – despite the fact that most astronomers were sure that no water or air exists there. In 1835, American writer Richard Adams Locke decided to mount an elaborate hoax, claiming that life had been observed on the Moon. In a series of articles in the *New York Sun*, he claimed that objects as small as 45 cm (18 inches) in diameter could be seen through the telescope of well-known British astronomer John Herschel. In a series of articles, Locke described a wealth of Selenite creatures, including a one-horned goat, a large, spherical amphibian and even apparently intelligent, winged, human-like creatures. The articles caused so much public interest that, for a while, the *New York Sun* became the highest-selling newspaper in the world. But during the second half of the century, more people realised that there was no way the Moon could be home to alien life. In 1865, the French science-fiction author Jules Verne wrote *De la Terre à la Lune* ['From the Earth to the Moon'], which included a discussion of

▲ French writer Jules Verne, who wrote many classic science-fiction adventure stories, including *From the Earth to the Moon*. Verne had no scientific training, but his painstaking research ensured that the scientific ideas contained in his books were as accurate as possible.

▲ 'New Discoveries on the Moon' – a print illustrating some of the discoveries reported in the *New York Sun* as part of the Great Moon Hoax of 1835. Selenite creatures such as these were supposedly observed on the Moon by British astronomer John Herschel.

William Herschel's 12-metre-long telescope, with its 120-cm-diameter metal mirror, built in Slough and completed in 1787. Using large telescopes like these, astronomers of the 18th and 19th centuries began to understand the true nature of space.

Selenites: '... if Selenites do exist, that race of beings assuredly must live without breathing, for – I warn you – there is not the smallest particle of air on the surface of the Moon'. By the end of the century, attention had turned to another location in space: Mars.

Martian Frenzy

The possibility of the existence of life on Mars has captured the public imagination to such a degree that, for many people, 'Martian' is synonymous with 'alien'. The main reason for the interest in Martian life is that Mars is the planet most like our own: it has an atmosphere, a solid surface, and polar ice caps (made mostly of frozen carbon dioxide) that grow and shrink with the Martian seasons. The similarity between the Earth and Mars was first noticed in 1783 by German-British astronomer William Herschel, who saw no reason why life should not exist there. Herschel had visions of clouds of water vapour, lakes and abundant vegetation. Equally, Martian gravity is not too dissimilar from Earth's and, while most other planets are too close to or too far from the Sun for life to be possible, Mars could be just hospitable enough to harbour life. Certainly many astronomers thought so in the 19th century: in the 1890s, the scientific interest in Mars reached the general public, and thus began decades of Martian frenzy.

Public fascination was fuelled in 1895 when American astronomer Percival Lowell published a book called *Mars*. One of its central themes was the Martian civilisation that Lowell supposed was living on the planet. In writing his book, Lowell drew on observations made by the Italian astronomer Giovanni Schiaparelli. In the 1870s and 1880s, Schiaparelli had become convinced that he could see a number of very long, straight lines on the surface of Mars. He described the lines as *canali*, the Italian word for channels. Lowell proposed that the channels were actual canals built by intelligent creatures living on the planet. Lowell also made his own observations of Mars: he found several spots on the surface, which he imaginatively described as oases. He envisaged a complex and technologically advanced Martian civilisation, which used irrigation to carry water from the Martian polar ice caps to each oasis.

While the mainstream scientific community was very sceptical about Lowell's ideas, science-fiction writers picked up on the public fascination with them. Perhaps the best known of many novels and magazine-based stories about Mars is *The War of the Worlds*, written in 1898 by British science-fiction writer H. G. Wells. In Wells' story, Earth is invaded by a fleet of war-mongering Martians. On Halloween in 1938 American actor and director Orson Welles broadcast a version of Wells' novel in the form of a realistic radio news bulletin. So strong was the belief in Martian civilisation that thousands of listeners were convinced that the broadcast was real and panicked. Despite mounting evidence that Mars is cold, dry and, unlike Earth, unprotected against the Sun's more harmful radiation, the interest in Mars as a home to intelligent life continued well into the space-age years of the latter 20th century.

The New Science of Aliens

Throughout the 20th century, scientific progress continued at a breathtaking pace, and scientists and the public alike remained fascinated by the possibility of alien life.

▲ Artist's impression of supposed channels of water on Mars, published in 1884 in *Worlds in the Sky*, by French astronomer Camille Flammarion. The existence of channels on Mars was first suggested by Italian astronomer Giovanni Schiapparelli. There are no such channels on Mars.

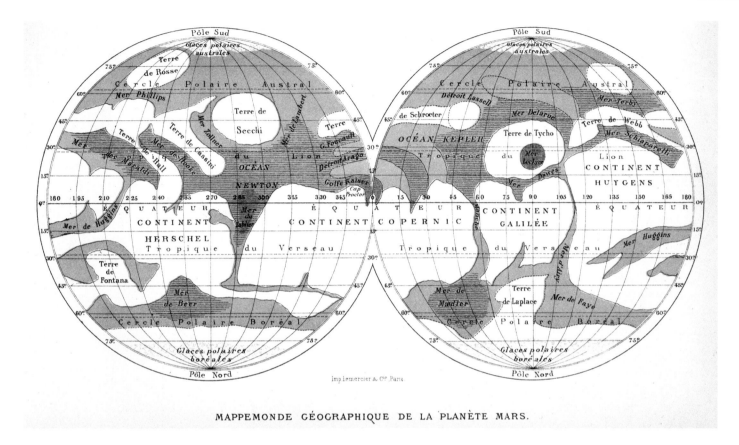

MAPPEMONDE GÉOGRAPHIQUE DE LA PLANÈTE MARS.

In astronomy, bigger and better telescopes and an increasing theoretical understanding of how stars and planets worked made the possibility of life somewhere in the Solar System seem less and less likely. The realisation, in the 1920s, that our Galaxy is one of countless millions brought a new sense of the scale of the Universe. In the 1970s space scientists sent robotic space probes out to explore the Solar System, and were thus were able to build up a picture of the planets in unprecedented detail. This knowledge seemed to reinforce the idea that the only place hospitable enough to harbour life is our own planet Earth. Some probes even carried experiments to test for signs of life. Most notable was the Viking mission to Mars during the 1970s, which included a small laboratory to test the soil. No unambiguous signs of Martian life were found.

▲ Top: 1892 map of the surface of Mars, showing what were perceived as oceans and continents. It is now known that Mars' surface is very dry, although water almost certainly flowed there millions of years ago. Above: Photograph of the surface of Mars, taken by the Viking 2 lander probe in 1976. The two Viking probes, launched in 1975, carried experiments to test the Martian soil.

The Science of Aliens

The 20th century also included many key biological discoveries, in particular the decoding of the molecule that is central to life on Earth: DNA (deoxyribonucleic acid). Biologists could finally understand and define life on a molecular level, and could begin to see how life probably developed by chance as a result of chemical reactions between naturally occurring, carbon-containing (organic) molecules. During the second half of the 20th century, astronomers were able to detect these organic molecules in comets in the Solar System, and more remotely in interstellar dust clouds, from which stars and planets form. This makes it likely that life has developed elsewhere. Moreover, the discovery of organisms adapted to living in extreme conditions here on Earth opens up the possibility that some kind of life might still be found somewhere in the Solar System.

Whether extraterrestrial life is ever found inside the Solar System or not, our scientific understanding of the processes of life make it seem very unlikely that we are completely alone in the Universe at large. Space science and biology are now so sophisticated that scientists are able to suggest what kind of life forms might one day be discovered on distant planets or their moons – perhaps even those in orbit around stars other than the Sun. But until the discovery of the first signs of extraterrestrial life, we must be content to wonder at the aliens of our imaginations. While scientists have recently been able to make educated guesses about what real alien life might be like, we begin our journey with those aliens that have been imagined by science fiction.

▲ Orson Welles, photographed during his famous radio broadcast of H. G. Wells' alien invasion story, *The War of the Worlds*, on Halloween, 1938.

Poster for the original film adaptation of *The War of the Worlds* (1953). A modern version, starring Tom Cruise (2005), has brought H. G. Wells' story into the 21st century.

▲ British actor Richard Wordsworth in the 1955 film *The Quatermass Xperiment*. Wordsworth plays Victor Carroon, an astronaut who undergoes a strange and dangerous metamorphosis after contact with extraterrestrial life forms.

Alien Stories

We have all had close encounters with aliens. Some of them are scary, others cute and cuddly; some are very similar to humans, others bizarre and unfamiliar; some are here to conquer us, others to teach or nurture us. These are, of course, fictional aliens found in science-fiction stories, films, television series and comic strips. The aliens of popular culture have also made their presence felt in adverts, computer games, hoaxes, urban myths and some yet-unsolved mysteries of alien abduction or UFO sightings.

Despite the claims of some people, there is no hard evidence that any real alien life forms have been to Earth. So, every alien we have ever 'met' is the product of the human imagination, rather than a result of evolution on a distant planet. Analysis of imagined aliens tells us more about our own hopes and fears than about extraterrestrial life forms that might really exist.

Aliens in Print

Extraterrestrials have been a staple of science-fiction stories since the end of the 19th century. Science fiction first became popular in Europe – particularly in France and Great Britain – but its influence quickly spread elsewhere. The US produced the most writers in the genre during the 20th century. Many early aliens were portrayed as evil or in some way threatening. The word 'alien' can have negative connotations – meaning strange or different – so perhaps this comes as no surprise. One recurrent idea was that alien races have superior intellectual and technological capabilities, at the expense of emotions and a sense of morality. In that case, they would be both a direct threat and a warning of what could happen to our own species. Almost all aliens came from planets in the Solar System – mostly from Mars, but also from Venus, Jupiter and Saturn. Although these early stories referred to scientific themes, most of them paid little attention to the theories of the day, and were closer to fantasy than to science fiction.

The Science of Aliens

▲ Not all early imagined aliens came from the Moon or Mars. This fanciful engraving, taken from Camille Flammarion's 1884 book, *Worlds in the Sky*, shows the imagined inhabitants of planet Venus.

Some science-fiction authors wrote books, but many of the early stories that featured aliens were found in 'pulp magazines', so-called because of the cheap wood pulp used to make the paper on which they were printed. The first pulp magazine devoted entirely to science fiction was *Amazing Stories,* a US publication first produced in 1926. Perhaps more important was *Astounding Stories*, which appeared in 1930. The editor of *Astounding Stories* from 1937 was John W. Campbell, who is considered by many to be the father of modern science fiction. Campbell set his authors very high standards in both scientific content and the quality of the writing. As science fiction gained respectability as serious literature, fictional aliens began to originate from outside the Solar System, and many were more imaginative creations than bug-eyed monsters and 'space people'.

Science fiction that is based on real theories of physics, chemistry and biology – rather than on the social sciences or simply on uninformed speculation – is called 'hard science fiction'. Among the most influential of the 'hard science-fiction' authors are Arthur C. Clarke, Robert A. Heinlein and Isaac Asimov. As science-fiction novels became 'harder', they also became rather esoteric, appealing only to readers with an insight into the scientific theories of the day. Science-fiction novels continue to have a large readership, but today it is in films and on television that most people come face-to-face with aliens.

Aliens on Film

A few early silent films included meetings with extraterrestrials, most notably, Georges Méliès' *Le Voyage*

Arthur C. Clarke, who wrote many important works of science fiction, including *2001: A Space Odyssey*. Like many other 'hard' science-fiction writers, he had a scientific background, receiving a degree in physics and mathematics from London University.

1950s' film posters, clockwise from top left: *Invaders from Mars* (1953), *This Island, Earth* (1955), *Invasion of the Saucer-Men* (1957), *Invasion of the Body Snatchers* (1955).

1950s' Science Fiction

The 1950s was the heyday of science-fiction films. While most depicted aliens as evil aggressors, coming to Earth to conquer or destroy humans, there were several notable exceptions to the 'conquering aliens in flying saucers' stereotype. *Forbidden Planet* (1956, above) stands out, not only because it is set on a distant planet far in the future, but perhaps also because it is a loose adaptation of Shakespeare's *The Tempest*. This atmospheric and thought-provoking film memorably features an all-purpose robot called Robby.

The Science of Aliens

dans la Lune ('A Trip to the Moon', known in the USA as 'A Trip to Mars', 1902) and *Aelita* (1924), a Russian film whose central character was the queen of Mars. The most notable of the few science-fiction films made during the 1930s was *Flash Gordon*, a series of films in which an unwitting human hero battles against Ming the Merciless, the evil emperor of the planet Mongo. But it was the 1950s when aliens really made their presence felt in popular culture, in a cluster of classic and not-so-classic feature films.

The science-fiction films of the 1950s often reflected post-war anxiety about nuclear weapons, and the growing Cold War tensions between the US and the USSR. The aliens portrayed in these films were often monsters bent on destroying the human race. Titles of 1950s' science-fiction films include *The Thing from Another World* (1951), *Robot Monster* (1953), *It Came from Outer Space* (1953), *It Conquered this World* (1956), *Not of this Earth* (1957), and *It! The Terror from Beyond Space* (1958). A film version of H. G. Wells' *The War of the Worlds* (1953) fitted into this genre perfectly.

Another strong influence on the films of the 1950s was the public interest in 'flying saucers'. For twenty years on the covers of pulp magazines these objects had been the vehicles of choice for alien visitors . But in 1947, UFOs hit the news headlines. Pilot Kenneth Arnold reported seeing several objects moving 'like a saucer would if you skimmed it across water' near Mount Rainier in Washington State. Heightened by the space race, the reality of space travel became established in the public consciousness. Echoing these events, film titles of the period included *The Flying Saucer* (1950) and *Earth versus the Flying Saucers* (1956).

During the 1960s, the wave of science-fiction films abated, making way for some influential television series featuring aliens. From 1963, British viewers were entertained by a time-travelling extraterrestrial in the series *Doctor Who*, and were frightened by his alien adversaries. Several episodes of the US series *The Outer Limits* dealt with extraterrestrial life, while the extremely popular *Star Trek* began in 1966. By the end of the decade, science fiction again hit the big screen. One film, *2001: A Space Odyssey* (1968) was groundbreaking for its big-budget special effects. The Moon landings and the launch of several planetary probes spawned renewed interest in space, and a new generation of Hollywood blockbuster films soon followed. The first of George Lucas' *Star Wars* films caused a sensation when it was released in 1977. In the same year, Steven Spielberg's *Close Encounters of the Third Kind* tapped into a surge of interest in alien visitation. The commercial success of these two films prompted Hollywood producers to consider making more science fiction.

Well-known science-fiction films since 1977 include the *Alien* quartet (from 1979), *E.T. the Extra-Terrestrial* (1982) and *Predator* (1987). The theme of visitation by aliens continues to be popular in films such as *Independence Day* (1996), in which warmongering aliens arrive at Earth in a huge flying saucer one quarter of the mass of the Moon; and *Signs* (2002) in which aliens communicate through crop circles (see pp. 119–20). Aliens have become so much a part of cinema culture that even several comedy films have been made. For example, Tim Burton's *Mars Attacks* (1996) is a hilarious spoof of 1950s' Earth-invasion films; *Men in Black* (1997) focuses on two secret agents who find

Toy dalek, manufactured in 1966. The daleks are human-sized alien enemies of The Doctor, himself an extraterrestrial, in the popular British television series *Doctor Who*. They are cyborgs: living creatures with mechanical body parts.

▲ Mr Spock (Leonard Nimoy) from the American television series Star Trek. Spock's mother is human, but his father is from an extraterrestrial race called Vulcans.

themselves caught up in an interstellar terrorist plot. The classic *War of the Worlds* was re-created by Steven Spielberg in a new film version released in 2005.

Alien Anthropomorphism

Most Hollywood science-fiction films are riddled with scientific inaccuracies of which most viewers are unaware. For example, spaceships are often seen firing laser beams across space – in reality, the beam itself would be invisible. Spacecraft engines can often be heard whizzing across space – and yet, in the near-vacuum of outer space, there is no sound at all. But there is one feature common to many science-fiction films that troubles not only the scientifically literate: why do so many aliens look like humans? More specifically, why are so many fictional aliens bipedal (two-legged), binaural (two-eared) and binocular (two-eyed)?

While some of the early aliens were bug-eyed monsters with death rays, most of them were 'space people'. In the formative days of science fiction, authors had less hard science on which to draw when designing their alien characters. The theory of evolution, which explains the development of species, was still new, and was by no means universally accepted. Furthermore, the biochemical basis of evolution was completely unknown, and little had been discovered about the stars and the planets. There was also, perhaps, a religious basis to anthropomorphism in science fiction: according to the Old Testament in the Bible, God created Man in his own image, so it was natural to assume that if he had created any other beings, they too would be in a similar image.

One justification for anthropomorphic aliens appeals to evolution itself. The idea is this: the senses and body shape we have are advantageous to us here on Earth, so if other life forms are living in similar environments, evolution may produce similar results. Another idea is that there was once a race similar to humans that existed elsewhere in the galaxy, spreading from planet to planet, and which changed only slightly as time passed. This is interesting but at odds with the observed lineage of humans, which stretches right back to the very beginning of life on Earth. Alternatively, aliens are represented as human-like reptiles, based on the belief that, had they not been wiped out 65 million years ago, the dinosaurs may have evolved into a globally dominant, intelligent species.

In some stories, very un-human aliens conveniently take on human form when they reach Earth. When aliens are portrayed in science-fiction films and television series, they may appear similar to humans for a very mundane reason: they have to be played by human actors. But the real root of the anthropomorphism of aliens in films has more to do with storytelling than with science or acting: giving fictional aliens human characteristics enables us to connect with them emotionally. This is also probably why so many supposed alien sightings feature 'little green men' or 'greys'. An alien that is similar, but different, from ourselves encourages us to reflect on our own humanity. Causing an emotional response is often the fundamental aim in designing most fictional aliens – humanoid or not.

Fear of the (Un)known

Throughout history, in legends and fables, we see many

The Science of Aliens

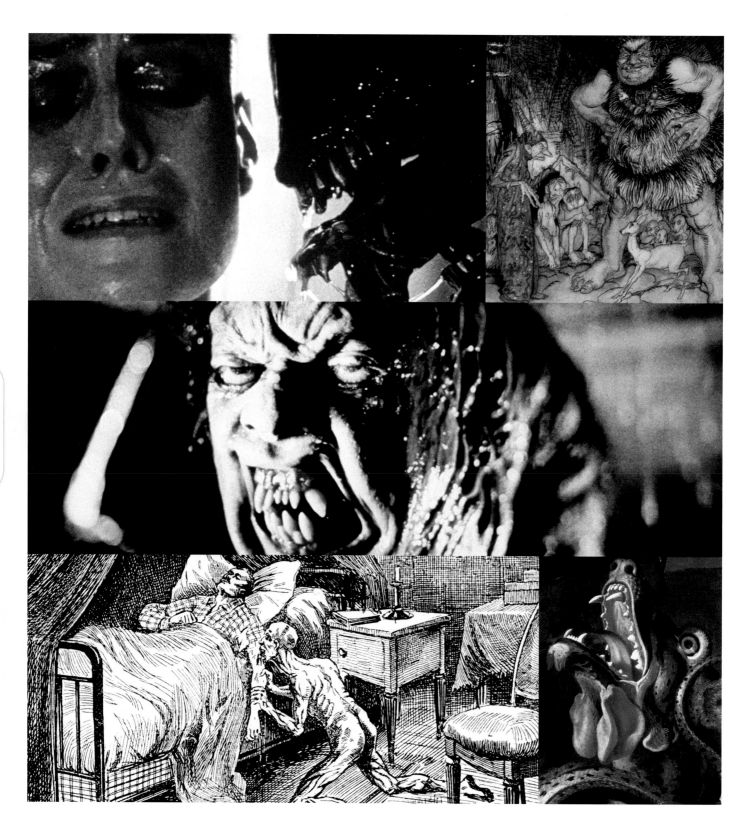

▲ Scary aliens and their mythical counterparts. Top: *Alien³* (1992); a giant called Galligantua from the story of *Jack the Giant Killer*. Middle: violent, shape-shifting alien in *The Thing* (1982). Bottom: vampire; the devil in the form of a dragon taken from Nicolas Poussin's (1594–1665) painting of Saint Margaret.

Gallery of Fear

In scary monster stories of any genre, humans are under threat. The aliens that provoke fear in readers or viewers of science fiction have much in common with frightening characters in myths, legends and fairy-tales. A flash of teeth dripping with saliva, animal instincts and evil intent are features that cause an immediate fight-or-flight reaction – an adrenaline rush.

The alien creature in the *Predator* films, shown here hunting down a New York police officer in *Predator 2* (1990), encapsulates all of these characteristics. A big-game hunter that travels through space to track down rare species, it has come to Earth to kill humans. It is large, like the giants of legends, it is monstrous like a dragon, and it has a taste for blood like a vampire.

▲ Yoda (voice by Frank Oz), a Jedi master from the *Star Wars* films. Far from evil and threatening, this cute and wise alien stands only 66 centimetres tall.

predecessors of evil aliens. Derived from the Greek 'daimon', which means 'supernatural being', the word 'demon' is now often used as a generic term for an evil creature. Originally, however, demons could be good or evil: they were simply personifications of the influences on a person's character. But the evil connotation prevailed in most cultures. In Judaism, there is a heirarchy of evil demons, whose prince, Satan, became identified with the Devil in the Christian faith. The Devil is a fallen angel, a malevolent spirit, who tempts humans into evil acts. Islam, too, recognises the importance of the Devil in this context, using his name, Iblis. The concept of demons has passed into everyday, secular language: people talk of 'casting out their demons' when they are keen to change part of their character with which they are unhappy.

Another popular fictional being that is often seen as evil is the dragon. Originally, the word dragon described any large, serpent-like creature, and was seen as neither good nor bad. In China, dragons are considered noble creatures. But Christianity saw the dragon as a creation of the devil, and many Christian works of art featured saints triumphing over them. As dragons passed from religion into folklore, their evil image – as something that must be slain – persisted. The dragon is the archetypal 'monster', often representative of the unknown – or alien – and as something to fear. Dragons and other monsters were often depicted in uncharted areas at the edge of old maps (although the well-known phrase 'Here be dragons' appears, in Latin, on only one).

Another archetypal monster, the vampire, originated in the Slavic countries early in the 18th century, but vampire stories were popular across many European regions.

The restless souls of criminals or people who have committed suicide, vampires are destined to live for eternity. They suck the blood of their human victims to survive, transforming them into vampires themselves. The vampire legend was immortalised in Bram Stoker's celebrated book *Dracula* (1897), as well as several films. Another mythical creature, the werewolf, is closely related to vampires. It too has featured in many influential tales of horror.

One common emotion experienced by viewers of science-fiction films is fear. In a sense, we are potentially fearful of all aliens, since by their very nature they are unknown to us. But the kind of fear exploited by writers and directors of science fiction is often more immediate and direct. In a well-made science-fiction horror story, we experience the 'fight-or-flight' response caused by a rush of adrenaline in the bloodstream. For many people, the physiological and psychological thrill of an adrenaline rush explains why they enjoy horror films. The classic era for science-fiction horror films was the 1950s. Many of the films of that decade drew on horror films of the 1920s and 1930s, for example *Nosferatu* (1922), *Frankenstein* (1931) and *The Mummy* (1932).

Perhaps the best recent horror film in the realm of science fiction is *Alien* (1979), directed by Ridley Scott and starring Sigourney Weaver as Ripley, a member of the crew of a mining ship in deep space. *Alien* contains many long passages of high suspense, in which the viewer expects imminent violent conflict between Ripley and an alien being that has hatched from an egg that was laid in another crew-member's body. When watching films such as *Alien*, viewers have empathy for the human characters, perhaps

The Science of Aliens

wondering what they would do in a similar situation. The tag line of the film was 'In space, no one can hear you scream'. But looking beyond the fear, the suspense and the empathy, there is something else that is common to almost all horror films: the sense of good triumphing over evil. We normally expect even a horror film to have a happy ending, despite what transpires as the plot unfolds.

By comparing the evil aliens of modern science fiction with evil creatures from mythology, folklore and fairy-tales, it is possible to understand the motivation of science-fiction writers and directors in creating their iconic aliens. In some ways, evil or monster aliens have more impact than traditional, folkloric creatures. Aliens are supposed to be part of the natural, not the supernatural, realm: they are therefore more credible in the modern world, where a scientific world view has made mythical beasts unbelievable. But not all science-fiction aliens are evil or monstrous.

Cute and Cuddly

Many fictional aliens can be described as 'cute'. Science-fiction authors and writers design these aliens to be lovable, again in order to provoke emotional attachment with their characters. Many of the features common to lovable aliens have also appeared on many dolls and cartoon characters: for example, Mickey Mouse, Barbie, Shrek, Bambi and the characters in Japanese *anime* cartoons.

Mickey Mouse is an interesting case. When he first appeared in 1928, he was something of a rogue, and his features were not as 'cute' as those of the modern-day Mickey, whose character and appearance are more lovable.

▲ The cuddly alien ET, from the film *ET: the Extra-Terrestrial*, directed by Steven Spielberg. Here, Gertie (Drew Barrymore) kisses ET just before he leaves to go back to his own planet.

Evolutionary biologist, Stephen J. Gould, made measurements of the relative size of Mickey's facial features as he 'evolved' through his cartoons. He found that the features of Mickey's modern appearance are similar to those that we associate with babies. These are the features that are also found in lovable aliens.

There is a number of respects in which babies' faces are very different from adults'. As children mature, their eyes stay almost exactly the same size: in other words, babies appear wide-eyed. The human jaw and nose are initially less pronounced, and the head itself is much larger, relative to the body size. Our emotional response to cute aliens can be traced back to our feelings for babies: we want to protect and nurture them. This response is elicited when we see animals or even inanimate objects that show

babyish characteristics. For example, compared to an adult dog, a young puppy has a much shorter snout, relatively much larger eyes and a larger head in proportion to its body. The classic example of this is the alien in the film *ET: the Extra-Terrestrial*. ET is small, gentle, snub-nosed, hairless and wide-eyed: he is also a long way from home. Yoda, from the *Star Wars* films, is another famously lovable alien creature, who has similar physical features.

Not all cute aliens share these characteristics. In an episode of *Star Trek* called 'The Trouble with Tribbles', the alien creatures are in the form of small, fluffy balls, which reproduce rapidly. In this case, the emotional response is probably related to the similarity of Tribbles to small, vulnerable animals. The Tribbles were, ironically, not cute at all, as they were attempting to invade the space ship,

▲ Frances Griffiths and the 'Cottingley Fairies', photographed by Frances' cousin, Elsie Wright, in 1917. Sadly a fake, this photo shows that the fascination with other-worldly beings is not restricted to aliens from outer space.

the USS Enterprise. The point of this was, no doubt, to challenge our emotional response (cute) with the reality of these creatures (threatening).

There are other aliens which, while not cute, are nevertheless not at all threatening or evil. They may be benevolent or simply vulnerable, and in either case, the viewers or readers identify with the characters as something they would like to become or something they are. The best-known bene-volent alien is Superman, who first appeared in a comic strip in *Action Comics* in 1938. The character of Superman, and his alter-ego Clark Kent, struck such a chord with the public that they went on to appear in radio broad-casts and blockbuster Hollywood films. The android robot C3PO in *Star Wars*, while highly intelligent, is

vulnerable as a result of his apparent incompetence.

As is true for scary aliens, cute, vulnerable and benevolent aliens also have their counterparts in myths, legends and fairy-tales. For example, fairies themselves are usually small, gentle and benevolent. In the Judeao-Christian-Islamic tradition, angels are benevolent beings that come to help or advise humans. There are several examples of aliens that have similar characteristics to angels: non-corporeal or glowing as a gentle light. In the film *Cocoon* (1985), aliens from a distant planet take on the form of humans for most of the film, but are revealed as glowing, angelic creatures towards the end. The aliens in *The Abyss* (1989) live under the sea, and glow with a translucent, luminescent quality.

▲ The Mekon, from Venus, was the arch-enemy of the comic strip hero, Dan Dare. He was very intelligent, as his extremely large forehead suggests.

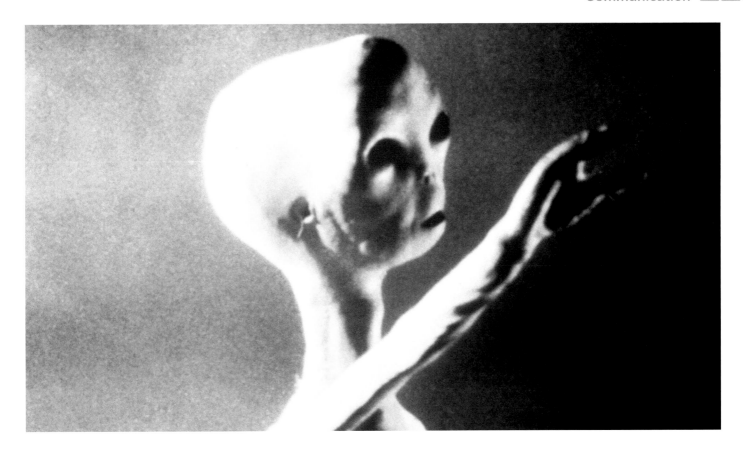

Magical Minds

Many of the fictional aliens that appear angelic or insubstantial possess an intelligence far superior to our own. Hyper-intelligence is one of the favourite characteristics given to fictional aliens. It makes sense to design aliens that are very intelligent, especially when they have travelled to Earth, a journey that would require technology far more advanced than our own. It is natural to wonder about what other, very intelligent civilisations might be like, for example those that are a million or a hundred million years older than our own. It encourages us to consider what humans might be like, and what kinds of lives they might live, far into the future. Perhaps future humans will be like the aliens that emerge from the spacecraft at the end of *Close Encounters of the Third Kind*: peaceful, gentle creatures. This serenity is apparent in many of the hyper-intelligent aliens depicted in science fiction.

Science-fiction writer Arthur C. Clarke wrote: 'Any sufficiently advanced technology is indistinguishable from magic', and science-fiction authors and directors make full use of this idea, allowing their hyper-intelligent aliens the freedom to read minds, to travel great distances in virtually no time, or to make objects appear and disappear at will. These technological abilities are reminiscent of sorcery, providing another link between science fiction and fairy-tales or folklore. The tradition of witches and wizards covers many hundreds of years and stretches across every continent. By being able to do things that ordinary humans cannot, witches and wizards set themselves apart from the rest of humanity – they are 'alien'. Mere mortals are both impressed and intimidated by hyper-intelligence, and super powers.

▲ One of the wise and peaceful aliens featured in the 1977 film *Close Encounters of the Third Kind*. Is this what humans will be like in the far future?

The Science of Aliens

As H. G. Wells' *The War of the Worlds* explored, the development of hyper-intelligence or very advanced technological capability can lead to a loss of morality and descent into outward aggression. More often, however, hyper-intelligent aliens are represented as wise. They may have come to Earth to scorn humanity for its failings. One of the best examples of this is found in the film *The Day the Earth Stood Still* (1951). In the plot, a spacecraft lands in front of the White House, in Washington D.C. A spaceman, Klaatu – a representative of an interplanetary federation – emerges from the craft to warn the human race that if they retain their aggressive ways when they venture into space, the federation will destroy planet Earth.

When a technologically advanced alien civilisation visits Earth, we humans are the vulnerable party. When we watch films like *The Day the Earth Stood Still*, we experience feelings of solidarity with our fellow humans, however much the aliens might have a point. When we read or watch science fiction about intelligent aliens coming to Earth, we may also experience a fear of being enslaved or subsumed into something to which we do not belong, thereby losing our identity. This was most apparent in American science-fiction films of the 1950s, many of which intentionally or unintentionally used aliens in this way as an allegory for the perceived threat from Communism.

During the 1950s – the early days of the Cold War – anti-communist hysteria was rife. There were many high-profile 'witch-hunts', for example, which were designed to root out communist sympathisers in the US. One film that seems to address or express this Cold War anxiety in America is

The invincible flying saucer in the 1951 film *The Day the Earth Stood Still*. The aliens it carried had an ultimatum to give to the aggressive human race.

40–41

▲ A rare glimpse of one of the warmongering Martians from the 1953 version of *War of the Worlds*.

The Science of Aliens

Invaders from Mars (1956). In the film, a boy sees a flying saucer land in the ground near his home. Anyone who goes to look for the saucer returns emotionless and with a small scar at the back of their neck. It takes the might of the American army to fight off the invading Martian (read 'communist') terror. Another film that exhibits Cold War paranoia is *The Invasion of the Body Snatchers* (1956). In this film, seed pods drifting through space from an alien world have landed on Earth. The aliens that emerge from the pods take on the physical appearance and the memories of their victims, but not their emotions.

It's Coming To See Us

It is easy to appreciate how the typical alien characters of popular science fiction have evolved through the context of storytelling, and with reference to their predecessors in folklore. But what about real people's claims of encounters with real aliens? What can we say about sightings of unidentified flying objects (UFOs), abductions and alien autopsies? Can they be explained in the same way? Or might they be genuine proof of the existence of extraterrestrial life, and proof also that aliens routinely visit Earth?

Within weeks of Kenneth Arnold's well-publicised 1947 sighting of objects that moved 'like saucers skimming on water', more than 800 people had contacted the US Air Force claiming to have seen flying saucers. The belief grew that these saucers were spacecraft from other planets, piloted by aliens. This belief is called the Extraterrestrial Hypothesis.

As a result of the UFO sightings of the late 1940s, the US Air Force set up the first of three projects to record and investigate UFO sightings. Some of the sightings seemed to be inexplicable, and so the Central Intelligence Agency (CIA) decided to bring together a panel of experts to assess any potential threat to national security. Hearing the news that government agencies were involved, the public became very interested in flying saucers. The last, and most famous, of the Air Force projects was Project Blue Book, which ran from 1952 until 1970. It recorded 12,618 UFO reports, of which all but 701 could be explained as illusions – reflections in windows, the planet Venus, aircraft lights and so on. By 1969, a committee that had been set up to report on the status of the project reported that there was no threat to the nation, and that there was no evidence that any of the unidentified sightings were attributable to extra-terrestrial vehicles.

The character of sightings and experiences relating to UFOs underwent gradual change from the 1950s to the end of the century. In the 1950s, reports almost exclusively described saucer- or cigar-shaped objects, or simply lights in the sky. Soon, there were those who began claiming that they had seen or even met the aliens in the spacecraft. The most famous of these 'contactees' was a man named George Adamski, who claimed he had met with human-looking aliens from Venus in the Mojave Desert, California. During the 1960s, people began to claim that they had been abducted by aliens. The first case of supposed alien abduction involved Barney and Betty Hill. In 1961 the Hills described how they had a close encounter with a cigar-shaped object with windows; the next thing they remembered was being on the road back home. They realised that there were two hours of their lives for which they had no memories of what had happened.

Under hypnosis in 1964, the Hills recounted being taken against their will, and operated on by aliens.

Barney and Betty Hill described their kidnappers as having large, olive-shaped eyes and grey skin, a description that has informed the popular archetype of modern aliens. Since the Hills' experience was made public, there have been about 120,000 other alien abductions reported worldwide. These sightings captured the public imagination, and fed back into science fiction. The aliens in *Close Encounters of the Third Kind* match very closely the iconic alien that developed during the 1960s and 1970s, often referred to as grey aliens, or 'greys'. During the 1970s, there were reports of cattle being mutilated near the sites of UFO sightings, and some people began to report that during abductions the aliens had implanted objects in their bodies.

Also during the 1970s, there was renewed interest in an incident that had begun some 30 years earlier. A few days before Kenneth Arnold's UFO sighting in 1947, a rancher in Roswell, New Mexico, had come across some strange debris on his land. After seeing the news report about Kenneth Arnold, the rancher went to the local Air Force base with the fragments. He announced that he had found 'one of them flying saucers'. At first, the Air Force publicly announced that a flying saucer had crashed, and it was in their possession. The news spread worldwide. After a few days, the Air Force declared that the debris was, in fact, from one of their weather balloons, and the event faded into obscurity for the next 30 years. Then, in 1978, two researchers found inconsistencies in the official version of events at Roswell. They became suspicious that there had been a cover-up. Further investigations led to new revelations, including the claim that bodies were recovered from the crash site. Just less than 20 years later, remarkable film footage was released that purported to show an autopsy carried out on one of the alien bodies recovered from Roswell. The Air Force maintains that the bodies were crash test dummies and that the autopsy is a fake. The crashed object, it claims, was a balloon designed to spy on Russia's nuclear programme.

Belief in the Extraterrestrial Hypothesis is widespread. Many people are convinced that, with so many sightings and other experiences concerning aliens and their vehicles, there can be no other explanation. There are many books, websites and organisations bent on uncovering government conspiracies. The film *Men in Black* was based on a conspiracy theory – the idea that there really are special agents who are in regular contact with aliens on earth, and whose job it is to keep the evidence from the public at large.

There are many different explanations for the lights and cigar- or saucer-shaped objects people see in the sky. Such explanations may not account for all of the reported sightings, but it is not necessary to explain away every sighting in order to disprove the Extraterrestrial Hypothesis. In fact, no amount of evidence can disprove it. On the other hand, a single, unambiguous piece of evidence would validate (at least some of) the sightings and experiences, for example a clear, genuine photograph that stands up to rigorous scientific scrutiny; a single alien implant that is truly alien; a real alien specimen.

Shedding some Light

To those who do not believe the Extraterrestrial Hypothesis, it seems reasonable to suggest that it is rooted in human psychology. Lights and objects in the sky can be mistaken

The Science of Aliens

▲ Top: alien abduction scene from *The Invaders* (1935), written by John W. Campbell under the pseudonym Don A. Stuart; print from William Allingham's 1886 book, *The Fairies*, illustrating fairy abduction. Bottom: still from the film *Invasion of the Body Snatchers* (1956).

Alien Abduction and the 'Greys'

Stories of abduction or possession have long been present in popular myth and fairy-tales. For example, there are many examples of stories, from many countries and across several centuries, which involve the abduction of children. Often, the children are made to drink a strange intoxicating liquid, and sometimes they return with strange bruises.

Abduction by aliens is a popular theme in science fiction. In *Invasion of the Body Snatchers*, aliens kill humans and take on their physical form. But more commonly, and especially in claims of genuine alien encounters, victims recount being taken physically for scientific examination.

The archetypal description of aliens in modern encounter claims arose in the 1960s, and was adopted into several films of the 1970s and 1980s. In contrast to the bug-eyed monsters common in early science fiction, modern 'greys' have large, olive-shaped eyes, sleek figures and grey skin.

A replica of one of the aliens that some people believe crashed at the Roswell Air Force base, New Mexico, in 1947. The replica is based on a film that seems to show a post-mortem examination of two aliens. This haunting figure forms part of an exhibition at the International UFO Museum and Research Center in Roswell.

for alien spacecraft only if the idea of an alien spacecraft already exists in the witnesses' minds. But psychology provides a deeper, cultural explanation for why the phenomenon has spread so widely. It is one of the features of the human brain to make sense of what it experiences. There are many documented experiments that demonstrate how fallible the human brain can be, especially under the influence of conscious or subconscious suggestion. At the beginning of the UFO phenomenon, suggestion existed in the form of a growing anticipation of space travel, and on the covers of pulp magazines from the previous 20 years.

There is another psychological basis on which to understand the UFO phenomenon: the desire for the aliens of science fiction to be real. As discussed above, we invest a good deal of emotional energy in strong characters that we love or hate. We associate fictional characters such as aliens with our own hopes and fears. This desire, reinforced as a shared experience, is enough to make people believe. In this way, the UFO phenomenon is often called a modern myth – a shared story that reflects what we believe or what we want to believe. Myths develop over time, from a few isolated occurrences into a consistent story. This is what has happened with UFOs.

There are cultural precedents to all of the features of the UFO phenomenon. For example, there was a similar wave of sightings – of what were called 'air ships' – in the US in 1896 and 1897. Encouraged by hoaxed or inaccurate news reports, and the fact that Mars had become a popular subject of science fiction, many people believed that the air ships were from Mars. One report even claimed that a mystery Martian had died and been buried somewhere in Michigan.

Interestingly, descriptions of the craft involved large propellers and flapping wings, rather than shiny metal discs with rocket propulsion.

Air-ship fever happened when real heavier-than-air flight was anticipated. In a similar way, flying saucers anticipated space flight and reflected an existing interest in extraterrestrial life. In both cases, there were credible witnesses, tremendous media interest and even supposed contactees. Abduction stories also have antecedents in fairy-tales and legends: fairies or elves were said to take children and replace them with their own offspring, who were deformed in some way or lacked any emotion. Such children were known as 'changelings'. The concept of demonic possession can be seen as a kind of abduction – of the soul.

It is possible to think about fictional aliens in very human terms: are they 'out to get us', are they 'lovable', are they 'more intelligent than us'? In fiction, it is important to develop characters like these, to which we can relate as human beings; and the aliens of fiction seem to have crossed over into myth.

But those who do not believe in flying-saucer sightings and alien abductions can nevertheless believe in the existence of alien life. And real science has advanced so far that we can begin to understand how real alien life might develop. Taken out of the emotional context of stories, and designed without the requirement of an emotional response, what might real aliens be like? In order to find out, it is necessary to examine what scientists understand about life here on Earth, and about the distant planets where real alien creatures might live.

▲ Drawing of a classic almond-eyed grey alien, taken from the film *Communion* (1989). In the film, Whitley Streiber (Christopher Walken) draws this picture under hypnosis, after he has been abducted by aliens.

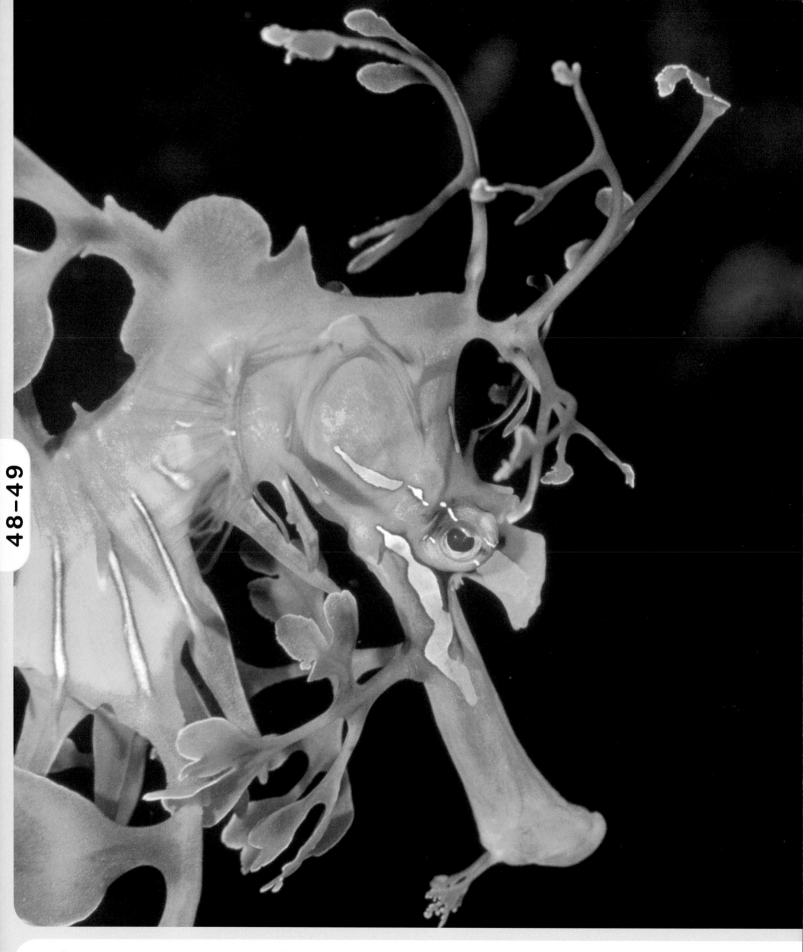

▲ There are some very alien-looking creatures right here on Earth. This is a leafy sea dragon, photographed off Kangaroo Island, Australia.

Aliens on Earth

The idea that our planet is being visited by aliens in flying saucers has waned in popularity since the 1980s, although there are still frequent UFO sightings and claims of government conspiracies and alien abductions. Mainstream thinking about extraterrestrial life has shifted towards a more 'down-to-Earth' approach. This is largely due to the creation of a new branch of science called astrobiology (also called exobiology and xenobiology), that has advanced greatly since the 1960s.

Astrobiologists combine what astronomers and astrophysicists know about space with what biologists know about life on Earth. The result is a sober, scientific investigation of the principles that might guide the development of life elsewhere. Sober it may be, but it is more exciting than the fantastic ideas presented in science fiction, and it is directing scientists in their search for the possible locations for life in space. Despite the efforts of space scientists and astrobiologists, Earth is still the only place in the Universe where we know that life has taken hold. What can science tell us about life on Earth, and how it began?

What is Life?

Philosophers have long pondered what makes some things alive, and everything else inanimate. During the 17th and 18th centuries, the emerging 'mechanistic' picture of the world proposed that living things are, fundamentally, the same as non-living matter: both are simply collections of atoms and molecules. This brought living matter within the realm of scientific study, but it did nothing to settle the question of what it is that makes something alive. Many scientists believed instead that living matter has some kind of 'vital force' that sets it apart. But that was no help either, since no one knew what that might be or where they might find it. Without a decent definition of what life actually is, how will we be sure if we ever find it elsewhere in the Universe?

The Science of Aliens

One approach to defining life is to list the processes that seem to be common to all living things: breathing, eating, growing, reproducing, metabolising (generating energy), excreting, moving, and responding to external stimuli. This is a circular definition: it simply states that a living thing is anything that has all the properties of living things. Furthermore, there are many exceptions to it. For example, on the basis of this list, there are many non-living things – such as cars or fire – which could be classed as living; conversely, there are some bacteria which would be considered non-living according to the list. Viruses fit hardly any of the criteria, although they are generally regarded as part of the living world. There are other schemes which attempt to pin down a definition for life, each with its own problems.

Perhaps the most robust definition of life – and the one of most interest to astrobiologists – relates to the theory of evolution by natural selection. Formulated during the 1840s (and published as *Origin of Species* in 1859) by British naturalist Charles Darwin, the theory of evolution explains how species develop over time. The theory assumes (correctly) that hereditary information carried in living things determines characteristics such as body shape and size. Darwin realised that this information, passed down through generations, was subject to random changes, or mutations. While most mutations are unfavourable, some produce distinct advantages in the battle for survival. The 'fittest' organisms thrive, and this can lead to changes within species and, gradually, to the development of entirely new species. Every living thing is a player in the story of evolution, so it makes sense to define a living thing as anything capable of taking part in evolution by natural

selection. This definition is useful to astrobiologists because it makes no mention of specific chemicals or biochemical processes, which might, of course, be very different on other planets and their moons.

All life on Earth depends on the chemical element carbon. The key to this element's importance in life is the ability of carbon atoms to bond with other carbon atoms, forming ring shapes, long chains and many different three-dimensional molecules. The role of this versatile element is vital (literally) in the formation of a complex molecule called DNA (deoxyribonucleic acid). DNA is a long molecule, and it exists in the nucleus of a living cell as two long strands twisted around each other. Small, carbon-based molecules called nucleotide bases form part of the twisting 'backbone' of the DNA molecule. The precise arrangement of these bases controls the manufacture of proteins, which build and regulate the cells of living organisms. Proteins are carbon-based molecules, too. Just as importantly, the totality of information carried by DNA – like a complete blueprint of the organism – is copied from generation to generation as organisms reproduce. The mutations required by Darwin's theory are 'mistakes' in this process of copying.

Many other molecules that take part in the fundamental processes of life are also based on carbon. The element is so important to living organisms that the study of the chemical reactions of carbon compounds is called 'organic chemistry'. Carbon atoms, and the molecules they form, are very versatile: while DNA itself may not exist on other planets, it seems likely that alien life will be carbon-based. One element whose atoms form chemical bonds in a similar way to carbon atoms is silicon. Some scientists (and science-fiction

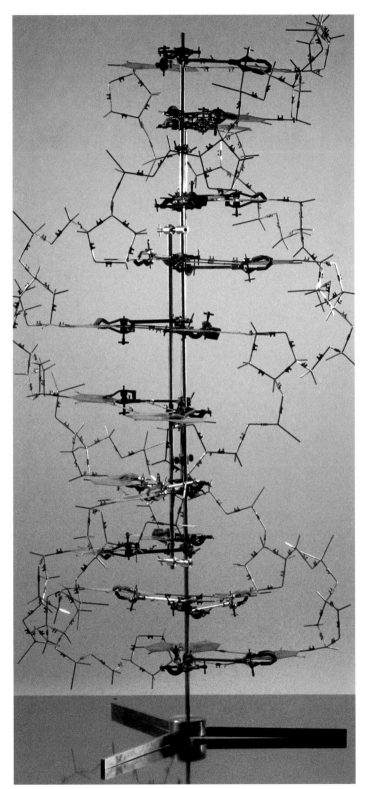

authors) have suggested that silicon-based alien life forms may exist. Both silicon and carbon are abundant throughout the Universe, but as silicon is less adaptable than carbon, the latter is the better bet.

The Tree of Life

Biological evolution produces an ever-diversifying 'tree' of life, with new species developing and some existing ones becoming extinct. Every branch of the tree is a separate species, which can fork off to form new branches. We humans are on a different branch of the evolutionary tree from chimpanzees, one of our closest relatives. But trace back along both branches about eight million years, and you find that chimpanzees and humans have a common ancestor, where the branches join. Look back even further – about 45 million years – and you find that humans, chimpanzees and lemurs have a common ancestor. Each branch of the tree is created when evolution provides new adaptations that create a new species.

As scientists investigate previously unexplored environments on Earth, they are discovering surprising ways in which living things have adapted to their surroundings. Some of the living things they have found are as alien to us as something we might expect to find on another planet. In the deepest parts of the ocean, for example, where hardly any sunlight penetrates, marine biologists have catalogued a bewildering array of weird and wonderful creatures managing to survive in conditions that are very inhospitable to humans.

By looking at some of the more unusual creatures that inhabit our planet, it is possible to see how evolution

The original laboratory model of the double-helix structure of DNA, constructed by the two scientists who worked out the molecule's structure: biophysicists John Watson and Francis Crick.

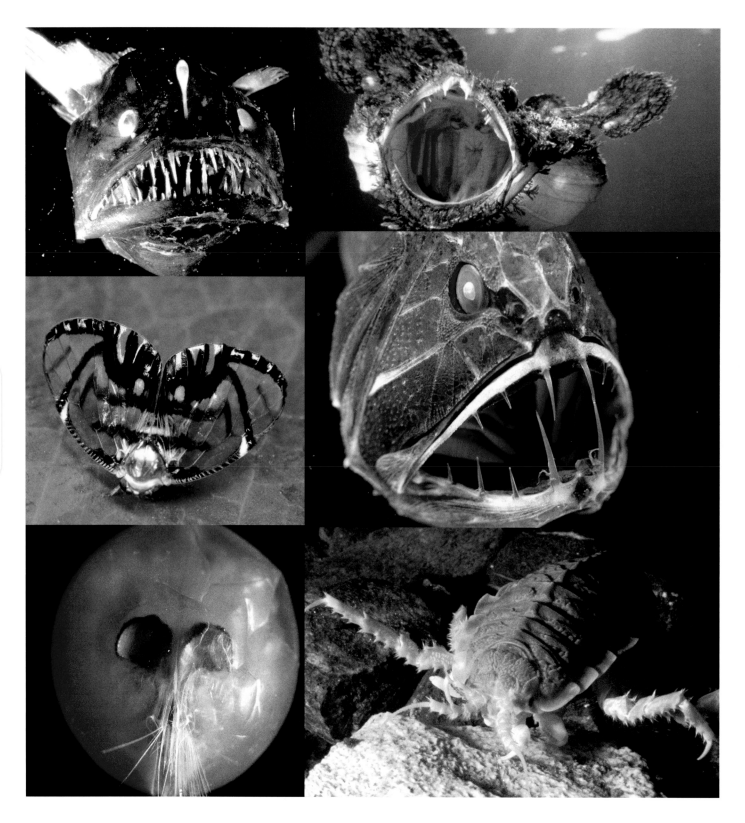

52-53

Clockwise from top left: Deep sea angler fish; angler fish attacking photographer; deep sea fangtooth; giant Antarctic isopod; giant ostracod; froghopper bug with wing markings that mimic a jumping spider.

Aliens on Earth

There are creatures here on Earth that would look very much at home in any imaginative science-fiction film. For example, cuttlefish (above) have well-developed brains, and are able to manipulate objects with their tentacles. They can change their body colour so that they blend into their surroundings, and their skin exhibits intricate and colourful visual displays when mating, as in this photograph.

Some of the most alien-looking creatures on our planet are found in the dark, high-pressure conditions of the deep ocean. The angler fish, for example, has a light-emitting organ that dangles near its fang-like teeth, attracting prey like bait at the end of a fishing line. Despite its ferocious appearance, it only grows to about 13 centimetres long.

Mimicry is one behaviour exhibited by many fictional aliens, and many terrestrial animals and plants have evolved this trait, too. The froghopper is an insect that is able to mimic spiders, in order to evade predators.

Scanning selectron micrograph of water bears. These remarkable 'terrestrial aliens' can survive intense radiation, very dry conditions, and temperatures from boiling water down to liquid nitrogen.

produces a wealth of different adaptations, each specific to the particular conditions in which the creature lives. Life on other planets will probably be just as varied, and we will certainly be thrilled and surprised at the adaptations we find, when and if we discover extraterrestrial life. But there is one aspect of evolution that might give us a clue about some features of life on another world. It is called 'convergent evolution'. Despite the enormous variety of living things on Earth, many important features have evolved many times – and completely independently – on unrelated branches of the evolutionary tree.

Evolutionary biologists know of many examples of convergent evolution. Birds have beaks, but so do octopuses and squid as well as certain (now extinct) dinosaurs. The common ancestor of these disparate creatures – way back in evolutionary history – did not have a beak. Similarly, there is a large number of organisms that have the ability to inject a venomous sting. They include certain insects, spiders, jellyfish and stinging nettles. Once again, the ancient common ancestor of these plants and animals did not have this feature. The best, and perhaps most important, examples of convergent evolution are vision and flight. Evolutionary biologists have identified nine distinct design types for eyes, including the 'camera' type found in humans. Across the nine types, some kind of vision has evolved independently at least 40 times. The means of flight has evolved at least five times.

One final interesting example of convergent evolution is sexual reproduction, where the DNA of two organisms of the same species combine to form a new organism. The reason why sexual reproduction, rather than asexual, has become the most successful means of reproduction on earth confounded biologists until recently. The advantage of sexual reproduction is that combining the DNA of two parents creates greater genetic variation. This enables species to adapt more quickly to changing conditions.

The fact that the same features spring up in many unrelated families of organisms does not mean they are somehow programmed into the process of evolution itself. Evolution is a random process: any advantageous feature that arises, through random mutation, is likely to remain and to develop further. Convergent evolution is simply the result of evolution producing several similar solutions to the same problems, quite by chance. Because convergent features evolve across a wide range of habitats, in very different plants and animals, astrobiologists believe that some of these features might also be found in organisms living on other planets. Perhaps one day in the future, someone from Earth will use their eyes to look into the eyes of a flying, stinging female creature from another planet. But what if life develops on a planet that is very different from Earth, for example, one that is very hot or very cold? Can evolution help to create organisms that can adapt to these conditions? The answer, it seems, is yes.

Living on the Edge

In the past thirty years or so, biologists have begun to discover a remarkable class of organisms that are adapted to survive the most extreme conditions on the planet. They are all single-celled micro-organisms, and collectively they have been called 'extremophiles'. Biologists have found evidence that the first life that appeared on Earth was similar to some

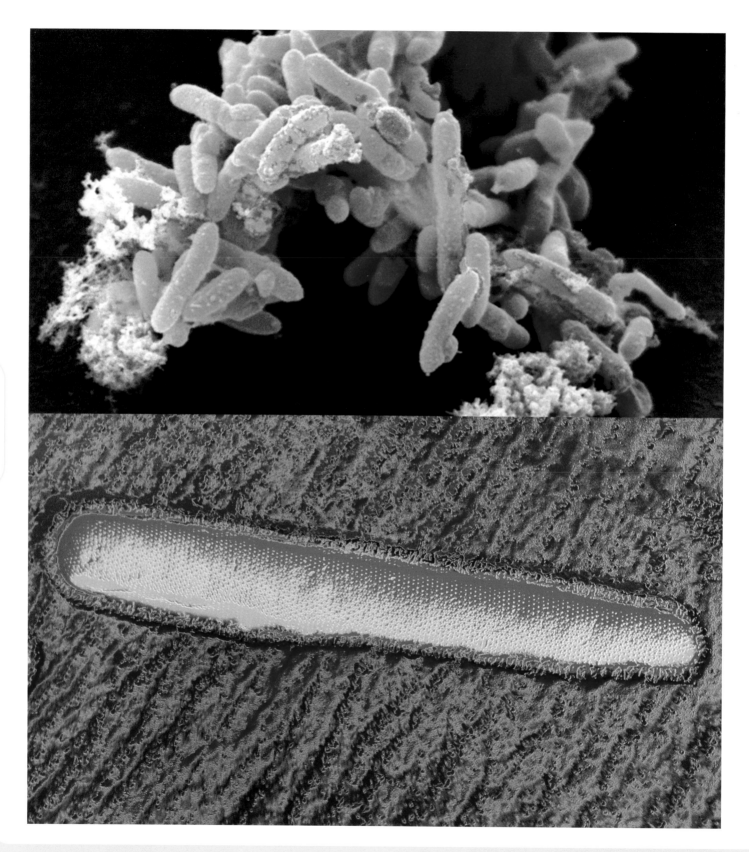

▲ Top: *Geobacter metallireducens*, a bacterium that digests uranium waste. Bottom: *Thermoproteus tenax*, a heat-loving bacterium. Opposite: an as-yet unclassified micro-organism, found in ice in Antarctica.

Life in Alien Environments

Extremophiles are micro-organisms that thrive in environments that would kill other creatures. The first extremophiles, heat-loving bacteria called thermophiles, were identified in the 1970s. Since then, some organisms have been discovered that live in strongly acidic conditions, some that can survive intense radioactivity and others that can enter into a state of suspended animation.

The orange bacterium shown above, given the nickname 'Klingon', was one of several bizarre, previously unknown species found in 400,000-year-old ice in Antarctica.

Scientists believe that extremophiles were the first organisms to appear on Earth, and something like them might also have been important in kick-starting life on other planets or their moons.

of these organisms. This makes sense, as conditions on the early Earth, when life first appeared, are known to have been hostile. The fact that life can thrive in extreme conditions is important to astrobiologists. It enables them to consider the existence of life on planets with hostile environments, which they may otherwise have overlooked.

The first extremophiles were discovered in hot springs, where water that has been heated underground by magma (liquid rock) comes to the surface. These organisms thrive at high temperatures, and are therefore called thermophiles. Most life depends upon enzymes – proteins and other molecules that regulate the functions of cells. Enzymes are normally very sensitive to temperature, and most cells stop working if the temperature is raised, whether they are single-celled organisms or part of a multicellular organism like a human being. But some thermophiles can only reproduce at temperatures above 90 degrees Celsius, and can survive temperatures well above boiling point for up to one hour.

Another family of extremophiles that is interesting to astrobiologists is called chemautotrophs. These strange organisms do not obtain their energy from the Sun, like plants, or by living off plants or animals, as most other living things do. Instead, they harvest energy from chemical reactions involving elements such as sulphur and iron. Other extremophiles can only survive in temperatures well below freezing; and there are those which thrive in concentrated acids or very alkaline conditions. One family of extremophiles can survive for long periods of time in a dormant state, without air or water. When Californian researchers examined a bee that had been perfectly preserved in an ancient sample of amber (fossilised tree sap) from the Dominican Republic, they found bacteria in its gut. They were able to remove and revive the bacteria, which had lain dormant inside the bee, inside the amber, for 25 million years.

The extremophiles are important, because they illustrate the fact that life can survive in harsh conditions. But if we are to find life somewhere in space, we need to understand how life can begin. Once again, there is only one example we can study: the origin of life on our own planet.

The Spark of Life

Every organism that has ever lived is part of the same evolutionary story. It has taken a very long time for evolution to produce living things as complex as human beings. The oldest known organisms lived about 3,800 million years (3.8 billion years) ago. They were discovered in sedimentary rocks in Greenland. These sedimentary rocks were laid down in water, suggesting that the early Earth can be regarded as a vast chemical laboratory, in which the complex organic compounds of life were produced. How did it all begin?

The first scientific theory of the origin of life emerged in the 1920s. According to the theory, important organic compounds such as amino acids (the building blocks of proteins) and sugars could have been produced spontaneously early in Earth's history from more basic ingredients that were known to be available. In 1953, chemists Stanley Miller and Harold Urey carried out a ground-breaking experiment to test the theory. They filled a flask with water and the compounds thought to be the main constituents of the atmosphere of the early Earth – methane, ammonia

▲ Five-hundred-million-year-old fossil of a trilobite. Now long extinct, trilobites were incredibly successful marine creatures until about 300 million years ago. Might anything like this exist on other, younger planets?

The Science of Aliens

A hydrothermal vent in the mid-Atlantic ridge, where molten rock, at temperatures of 360 degrees Celsius, meets seawater. Could terrestrial life – or extraterrestrial life – have begun in vents like these?

and hydrogen. Using energy supplied by an electric spark over many hours, to simulate lightning, amino acids and sugars were indeed produced. Since then, similar experiments have yielded other important organic chemicals, including nucleotide bases (the constituents of DNA). Exciting though these results are, there are problems. For example, it is thought that the early Earth's atmosphere would have provided no protection against intense ultraviolet radiation from the Sun, which might have broken down any organic molecules. There is also dispute over the exact composition of the Earth's atmosphere at the time. Nonetheless, experiments like these do demonstrate the possibility that life may have arisen spontaneously as a result of some kind of 'chemical' evolution.

Another problem with experiments such as these is that there is a significant difference between the basic organic molecules produced and the complex, self-replicating ones involved in even the most primitive living things. One idea that might explain the production of the more complex chemicals involves underwater structures called 'black smokers', which look like chimneys. These structures – more properly called hydrothermal vents – are like volcanoes or hot springs on the ocean floor. They occur where magma from beneath the Earth's crust meets the ocean. In a theory put forward in 2002, biologists William Martin and Michael Russell suggest that tiny compartments within the black smokers might act in the same way as biological cells, and provide just the right environment for building complex organic molecules. Life on Earth may have begun in such hot, dark conditions at the bottom of the ocean. One of the exciting things about Martin and Russell's theory is

that it means that life could arise on any volcanic planet with liquid water.

Long before hydrothermal vents had been discovered, and decades before Miller and Urey simulated the early Earth in a flask, Swedish chemist Svante Arrhenius had his own idea about how life began on Earth. In 1908, Arrhenius suggested that bacteria might be spread about the Universe, from one planetary system to another, seeding new life wherever they land. He called his theory 'panspermia'. Despite the fact that it still begs the question of how life actually began, panspermia is an interesting idea. If the theory is correct, then Earth may be just one niche in a biosphere that encompasses much of the Galaxy. In that case, there could be pools of carbon-based life around many different stars, all with a common origin. No extremophiles have yet been discovered that can survive intense radiation, such as they would experience in space. This is one reason why astrobiologists reject the panspermia theory: the 'seed' bacteria would surely not survive the journey from one star system to another. However, there are modern theories along similar lines.

There is a chance that the complex organic chemicals necessary for the beginning of life arrived on Earth, ready-made, from space. Various organic molecules are known to exist in comets and asteroids, and they have even been found in meteorites that have survived the journey to Earth. In the early history of our planet, large meteorite impacts were much more common than today. Perhaps even more exciting, astronomers using radio telescopes have been able to detect organic compounds in interstellar dust clouds. These regions of space, between the stars, contain tiny

Clockwise from top: Long-eared bat; Archaeopteryx; common darter dragonfly; ruby throated hummingbird feeding. Opposite: red kite in flight.

Convergent Evolution: Flight

Evolution produces a fantastic diversity of living things on Earth, but there are many features that have arisen independently several times in unrelated species. It seems likely that some or all of these features will exist in organisms on other planets, where evolution will be inevitable if life is to survive.

One such 'convergent' feature is flight. Insects were the first to take to the air in powered flight. The first flying insects evolved at least 300 million years ago. Next were flying dinosaurs – the first vertebrates to develop wings. The most famous flying reptile, the pterosaur, appeared 225 million years ago.

The oldest known fossilised animal that is generally considered a bird is Archaeopteryx. Fossilised specimens up to the size of a chicken have been found, which lived about 150 million years ago. Birds themselves are remarkably successful and diverse. Bats – the only truly flying mammals – first appeared about 55 million years ago.

The Science of Aliens

granules upon which atoms of carbon and other elements come together and react. It is from interstellar dust and gas that stars and planets form. The chemicals of life as we know it are certainly very common in the Universe. Our galaxy – perhaps all galaxies – may be teeming with carbon-based life.

Our Cosmic Neighbourhood

It will be a very long time before humans or robots from Earth manage to explore planets around other stars. But the existence of extremophiles on Earth has given astro-biologists renewed hope that some kind of life might exist somewhere in the Solar System. There are a few locations where space scientists are focussing their search.

The planets that are closer to the Sun than Earth – Mercury and Venus – receive intense radiation from the Sun, and their surface temperatures can be extremely high. Mercury has a strange orbit, in which the 'day' is longer than the 'year'. On Mercury, the Sun sets 176 days after it rises. During this long Mercurial day, the surface temperature rises to a scorchingly hot 450 degrees Celsius. During the equally

long night, temperatures plummet to minus 187 degrees Celsius. The high temperatures on the sunlit side, coupled with the fact that Mercury is small, means that the planet has virtually no atmosphere, and certainly no liquid water. Mercury is as dry and almost as airless as the Moon. Venus, on the other hand, is blanketed by a very dense atmo-sphere. As well as clouds of concentrated sulphuric acid, there is a great deal of carbon dioxide in the atmosphere, and Venus has experienced a runaway greenhouse effect. The average temperature at the surface of Venus is a searing 464 degrees Celsius.

Earth orbits at just the right distance from the Sun for carbon-based life forms. Surface temperatures are neither too hot nor too cold for liquid water to exist. However there is a possibility that Earth could go the same way of Venus with a greenhouse effect increasing the surface temperature to more than 400 degrees Celsius. But, for now, life thrives here. The next planet away from the Sun is Mars, long the favourite of science-fiction authors. Astronomers know that, long ago, planet Mars was warmer than it is today, and that the Martian atmosphere was more dense. Liquid water almost certainly flowed across the surface, and it may still exist deep underground. There is also convincing evidence of ancient Martian glaciers. It is quite possible that primitive life once existed on Mars (though no definite evidence has yet been discovered). If it did, perhaps there might be some remnants of it today: maybe some kind of extremophile living under the surface in a long-dormant state.

In 1996, Mars grabbed the public's attention, when NASA (the US National Aeronautics and Space Administration)

Today, planet Mars is a cold, dry and dusty place, unlikely to harbour life, as this Viking 2 photograph shows. But long ago, conditions were different; perhaps life once thrived here.

announced the discovery of possible signs of life in a meteorite that had come from Mars. The meteorite, known as ALH84001, was found in Antarctica in 1984. The rock would have been ejected from its home planet as a result of an energetic meteorite impact; scientists have amassed more than thirty meteorites that are known to have come from Mars. It has been estimated that the rock left its planet several million years ago, and that it landed on ours several thousand years ago. Researchers examining the meteorite found what look like fossilised bacteria. Inside the rock they also found organic chemicals that are produced when bacteria decay, and it seemed like the rock had, in the distant past, been infiltrated by liquid water. The excitement faded when many scientists argued against the claims, but to this day, the jury is out on whether ALH84001

really is the first-ever direct evidence of alien life to be found. Whatever the truth, scientists tend to think that today there is no life to be found on Mars.

The two huge planets, Jupiter and Saturn, are so far from the Sun that they are much too cold for liquid water to exist. Saturn orbits ten times as far away from the Sun as Earth. These two planets, together with Uranus and Neptune, are 'gas giants'. At the core of each is probably rock, surrounded by a strange 'metallic' liquid form of hydrogen, but the rest is made of gas and droplets of liquid. There is no 'surface' as such, and the cloud-top temperature is well below freezing. Temperatures increase closer to the core – particularly in the case of Jupiter – but it is highly unlikely that any form of life can exist there. The attention of astrobiologists is presently turned to certain moons of

▲ A photomontage showing planet Jupiter (centre, top), with its four largest moons (left to right: Io, Europa, Ganymede and Callisto). Might some form of life exist on one of these moons?

The Science of Aliens

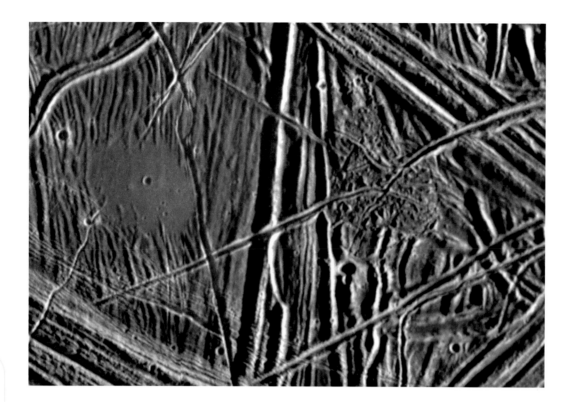

Jupiter and Saturn. Although these objects, like their parent planets, are well outside what is normally considered the habitable zone, they are special cases. Two moons in particular have encouraged a good deal of speculation: Jupiter's moon, Europa, and Saturn's moon, Titan.

Of the more than 60 moons known to orbit the massive planet Jupiter, Europa is the most tantalising. Discovered in 1610 by the Italian astronomer and mathematician Galileo Galilei, Europa is a small, round, rocky object that orbits its parent planet once every 84 hours. All Jupiter's moons hold some water, but most of it is frozen. This is true at Europa's surface, whose average temperature is minus 145 degrees Celsius. The resulting layer of ice reflects most light that hits it: if our Moon was covered in ice in this way, a moonlit night would appear more like twilight. From the outside, Europa appears very cold indeed, but it is heated from within. There are two sources of Europa's internal heating:

first, the gravitational forces of Jupiter and the other Jovian moons constantly stretch and squash Europa; second, radioactive elements around its centre decay, thus producing heat. Space scientists believe that Europa's internal heating leads to volcanic activity underneath the layer of ice.

The icy surface of Europa has patterns on it very similar to those observed in pack ice floating in the Arctic seas on Earth. Certainly the smooth icy surface has been observed to move, as glaciers and pack ice do on Earth. There are also lines across the surface, which are thought to be the result of warmer water welling up to the surface and freezing immediately – like a volcano of water. The emerging picture of Europa is of a moon whose internal heating systems melt some of the ice on the surface. Beneath the outermost layer of ice, there may be vast warm oceans of water, or at least slushy, half-melted ice. The exciting thing about Europa for astrobiologists is that

▲ Close-up photograph of the ridged, icy surface of Jupiter's moon, Europa. The smooth, circular region (about 3.2 kilometres in diameter) is the result of an upwelling of liquid from beneath the icy crust.

underwater volcanoes in an ice-covered ocean might be very close to the conditions on Earth when life began. Recent results from NASA's Galileo probe suggest that the layer of ice on Europa may be as little as a few kilometres thick. Another of Jupiter's moons which has created a stir is Callisto. As in the case of Europa, astronomers believe that Callisto may have a vast ocean of liquid water underneath an icy crust. Callisto's internal heating source is not as powerful as Europa's, but astrobiologists will certainly be keen to study it further.

Saturn has more than 30 moons. The largest, Titan, was discovered in 1655 by Dutch physicist and astronomer Christian Huygens. It provides what is perhaps the closest approximation to the composition of Earth's early atmosphere. Titan's atmosphere is mostly nitrogen and argon, but there are small amounts of carbon dioxide, methane, water and a number of interesting organic chemicals which form a dense smog. One of the compounds in the smog of organic chemicals is hydrogen cyanide, which some scientists believe might have been important in the development of life early in Earth's history. The organic chemicals are produced when ultraviolet radiation from the Sun breaks down methane high in the atmosphere.

From measurements of Titan's density, astronomers can tell that it is composed of roughly half rock and half water. Apart from the water vapour in the atmosphere, all the water is in the form of ice, since the average surface temperature is about minus 180 degrees Celsius. In January 2005, a probe called Huygens (part of the joint NASA and ESA [European Space Agency] Cassini-Huygens mission) landed on Titan. On its way down to the surface, Huygens captured images that show rivers of liquid flowing towards some kind of shoreline. In June 2005, the other part of the mission – the orbiting Cassini probe – captured amazing images which show what seems to be like a large lake on Titan. But any oceans or lakes on Titan are made of liquid methane, not water. The probe fell onto damp clay, and sent back remarkable pictures from the surface. Titan is a very cold, hostile place. The chances of finding any kind of life there are remote. But astrobiologists are excited by the organic chemical reactions that take place in its atmosphere. They see it as a golden opportunity to study the kind of reactions that, long ago, may have been responsible for the beginning of life on Earth.

Three different view of Saturn's moon, Titan, taken by the Cassini space probe. The image on the left shows what Titan would look like to the human eye. The reddish-brown haze is a photochemical smog rich in complex organic chemicals. The other two photographs are taken using special filters that allow the probe to 'see' features on the surface.

Clockwise from top left: nautilus – a small mollusc; caterpillar of the oleander hawk moth with false eye spots; squid; flatworm on a leaf. Opposite: horsefly with compound eyes.

Convergent Evolution: Eyes

Evolutionary biologists have recorded between 40 and 60 examples of vision that have arisen independently, in unrelated species. There are at least nine quite distinct designs of image-forming eye. They include the camera-type eye, which has a lens and forms good images, the multi-lensed compound eye, and pinhole eyes found in nautilus and some types of worm.

The camera-type eye is found in all vertebrates, some marine invertebrates such as octopus and squid, and even in some spiders. Compound eyes are found in a large number of arthropods, including many insects and some crustaceans. They consist of up to 1,000 tiny tubes, with a rigid lens at one end and light-sensitive cells at the other.

Vision provides such an advantage that it is very unlikely it will not have evolved in creatures on other inhabited worlds, should they exist.

The Science of Aliens

Further still from the Sun are the orbits of the other two gas giants, Uranus and Neptune, and then many smaller, rocky objects that are collectively known as 'plutinos'. The biggest of the plutinos is Pluto, which is normally considered a planet in its own right. The chances of finding any life on Uranus, Neptune or the plutinos seems extremely unlikely, so dark, cold and desolate is the Solar System at such distances from the Sun. Similar to the plutinos are asteroids. Hundreds of thousands of these chunks of rock orbit in a 'belt' between the orbits of Mars and Jupiter. The largest known asteroid is just 933 kilometres in diameter. A small proportion of asteroids contain organic chemicals and water frozen as permafrost beneath their surfaces, but none has more than an extremely tenuous atmosphere. The same goes for comets, although they have much more water (comets are often called 'dirty snowballs'). Life may have been kick-started by organic chemicals from asteroids and comets crashing down onto the surface of Earth early in its history; but there is really no chance that any life has ever existed on asteroids and comets themselves.

The locations where life – or even evidence of ancient life – might be found are very few and far between in our Solar System. In the next few decades, space scientists may find primitive life on one of the moons of Jupiter or Saturn. That would be an amazing discovery. But we live on the one planet at a favourable distance from the Sun where a great diversity of life can develop. The Universe has many, many stars other than our Sun. What can space scientists and astrobiologists say about the chances of finding life on distant planets?

Looking for Planets

During the 20th century, astrophysicists worked out in impressive detail how the Sun and the planets formed. The Solar System was born from a huge, cold cloud of gas and dust called a giant molecular cloud. There seemed to be no reason why the same process should not occur routinely elsewhere in our galaxy and beyond. And yet, until the 1990s, not one 'extrasolar' system of planets had been found. This was because stars are so far away that even through the most powerful telescopes, they remain as little more than points of light. Planets are much smaller than stars, and they do not shine – we can only see the planets of the Solar System because they reflect sunlight. So how do astronomers go about finding extrasolar planets, and begin working out whether conditions there might be favourable for the development of life?

Astronomers have had to design cunning ways to search for planets outside our Solar System. One method is based on a simple idea: as long as we are viewing a planetary system side-on, the planet must pass between the star and our telescopes here on Earth (or those in orbit, like the Hubble Space Telescope) once per orbit. As it does so, it will block a small amount of the star's light. So, measuring a star's brightness over a period of time would reveal regular dips, as the star is partially eclipsed by its attendant planet. By measuring the duration of each dimming, the orbital speed of the planet can be worked out; and from the magnitude of the dimming, astronomers can estimate the size of the planet. Several planets have been discovered using this photometric (light-measuring) approach.

By far the most successful technique for finding extra-solar planets is called the radial velocity method. It depends upon the gravitational interaction between a planet and its star. All objects exert gravitational forces on one other, tending to pull everything together: the fact that the planet is moving fast enough saves it from plummeting into the star. The planet pulls on its star with exactly the same strength as the star pulls on the planet. If the two objects had the same mass, they would both orbit around a point halfway between them, as if they were two dancers whirling around holding hands. They do both orbit the same point but, because of the difference in their masses, that point is inside the star itself or just outside it, resulting in a slight wobble in the star. Relative to its overall motion through space, the star appears to come towards us and move away from us over the same period as it takes for the planet to orbit. The heavier the planet, the greater the wobble. The closer the planet to the star, the faster the wobble. Early efforts to detect the wobble – by comparing photographs of a star taken at different times – proved ineffective. So astronomers now use a more indirect method: fortunately, they can measure changes in a star's speed to an accuracy of about one metre per second (3.6 kilometres per hour).

There are several other methods for detecting extra-solar planets: at the time of writing the total number of planets detected numbers around 150. Early in 2005, several astronomers released actual images of extra-solar planets. All of these planets are very large and very far away from their star. While there may conceivably be life on a gas giant or one of its moons, astronomers and astrobiologists would dearly love to find many more planets similar to our own. It is on Earth-like planets that life is most likely to develop.

Several missions and research programmes are planned for the coming decades which will find more, and smaller, planets, around a wide range of different types of star. Searches for planets around other stars were already under way when, in 1992, remarkable images from the Hubble Space Telescope showed in unprecedented detail a wealth of star formation in an interstellar cloud of gas and dust. Around many of the stars were disks of material showing planetary systems in the making.

Stars and planets are being made all the time, but stars do not last forever. Most stars shine for thousands of millions of years, before exploding as supernovae or simply cooling down slowly over millions of years. The chemicals that ultimately gave rise to life here on Earth were all contained in the interstellar gas cloud from which the Solar System was formed. Primitive life began on Earth only a few hundred million years after the Earth formed. Other planetary systems are forming all the time from clouds of gas and dust composed of a similar mixture of elements and compounds as the ones that made the Solar System. With so many stars in our galaxy, and so many other galaxies, it seems reasonable to suppose that carbon-based life has developed many times. Where it develops spontaneously life will certainly advance according to evolution. If and when space scientists do come across real examples of alien life, it will be well-adapted to its environment, otherwise it would not have survived. With all of the knowledge and understanding they have gained, space scientists, evolutionary biologists and astrobiologists can start to make educated guesses about how extraterrestrial life might look and behave in a range of very different environments.

▲ Top: spiral galaxy NGC 4414. Bottom: artwork of NASA's Cassini space probe releasing the Huygens probe down to Saturn's moon Titan – Saturn is visible in the background; Earth from space, taken by the crew of Apollo 17.

Our Place in Space

Our Sun is a star in the suburbs of a spiral galaxy similar to the one shown opposite. Our Galaxy contains around 400,000 million stars, and light takes 100,000 years to travel across it. The nearest full-sized galaxy beyond our own is 2.2 million light years away.

Space is vast. When discussing a distance in space, it makes sense to speak in terms of the time it takes for light to travel that far. Light takes about eight minutes to travel the 150 million kilometres from the Sun to our beautiful planet. To date, human beings have travelled no further than the Moon: light takes just over a second to make that journey. Robot space probes have explored far more distant objects in the Solar System. Saturn and its moons, for example, are never less than one light hour from Earth.

The true-colour image above was taken by the Hubble Space Telescope over ten nights in December 1995. It is the deepest view of space ever recorded, and shows the galaxies that lie in a tiny fraction of the sky. The light from the most distant ones travelled for around 10,000 million years before entering the telescope.

CARBON LIFE

FLEXIBILITY : HIGH
REACTIVITY : HIGH
ATOMIC NUMBER : 6
ATOMIC WEIGHT : 12.011
ATOMIC RADIUS : 77
IONISATION ENERGY : 11.2603
ELECTRO-NEGATIVITY : 2.5
ISOTOPES : ^{12}C ^{13}C ^{14}C
ELECTRON CONFIGURATION : [He] $2s^2$ $2p^2$
DENSITY : 2.25 G PER CM^3 / 0.08 LB PER IN^3

SOLAR PLANETS

RELATED DATA ◀ 1 2 3 4

PLAY ▶ PAUSE ‖

INTELLIGENCE   CARBON LIFE

▲ Working with the latest theories about life on Earth and sophisticated computer models, a team of scientists has imagined realistic life forms on two worlds in orbit around distant stars.

Alien Imagination

With several current and future projects aimed at finding extra-solar planets, the number of known worlds is set to increase dramatically. Some of these worlds might have conditions that favour the development of life – but which ones? Using sophisticated computer modelling, scientists have recently shown that planets and their moons around stars previously believed unfavourable might be perfect locations for life after all. They have even made educated guesses about what life might be like there.

One of the most important planet-finding missions to be launched in the coming years is the Terrestrial Planet Finder. As its name suggests, this mission aims to find planets that are similar to Earth: most extrasolar planets discovered to date are gas giants such as Jupiter and Saturn. The Terrestrial Planet Finder will be one of the most important projects funded by NASA (the US National Aeronautics and Space Administration) over the next 20 years. It will concentrate on searching for Earth-like planets within their star's habitable zone (where liquid water can exist). The mission will consist of two parts, both of which will be launched into space, beyond our planet's turbulent atmosphere. The first is an optical coronagraph, a powerful telescope that contains a disc which will prevent light from reaching the centre of the field of view. When the telescope is observing a star, the star's light will not enter the telescope: this will leave any orbiting planets visible. The main aim of this part of the mission is simply to find more planets; it should be launched in 2015.

The second part of the Terrestrial Planet Finder mission, to be launched sometime before 2020, is perhaps more exciting. It is called an infra-red interferometer, and will be composed of several individual telescopes flying in formation. By combining results from the individual telescopes, astronomers will be able to study distant planets in unprecedented detail. These telescopes will detect infra-red radiation, not visible light. Studying the infra-red radiation picked up from a planet enables astronomers to determine what gases fill that planet's atmosphere.

The Science of Aliens

The gases they detect should provide important clues to whether life exists there: biological processes produce different gases from non-living processes. In particular, scientists will be looking for high levels of oxygen, carbon dioxide and methane – the signature of life as we know it.

The Terrestrial Planet Finder will study planets within 50 light years from the Solar System – practically next door in cosmic terms. It will observe planets around a variety of different types of star. Which type of star is likely to harbour life-bearing planets? As life thrives on Earth, it is natural to suppose that stars like the Sun would provide the most favourable neighbourhoods for planets where life might develop. Such stars – called 'yellow dwarf' stars – are indeed good candidates for supporting life-bearing planets. They are stable and relatively long-lived, giving life a good chance to take hold. Unfortunately, yellow dwarf stars are relatively rare in our galaxy. Far more common are smaller, cooler stars called 'red dwarfs', which account for three out of every four stars. Red dwarfs are cooler than the Sun, so only a planet orbiting very close to such a star will lie within its habitable zone.

A strange thing happens to a planet that orbits very close to a star. The gravitational influence of the star slows its rotation to a speed where one hemisphere of the planet is always facing the star. On such planets, there is no regular cycle of light and dark, like day and night here on Earth. Instead, one side suffers an extremely cold, eternal night, while the other side experiences a never-ending day. From the light side of a planet like this, the red dwarf 'sun' always remains in the same position in a sky filled with unchanging daylight. Another feature of red dwarfs is that they

frequently flare-up, sending bursts of intense, harmful ultraviolet and X-ray radiation out into space. These flares can be very harmful to life, breaking apart carbon bonds that are so important in complex organic chemicals. The flares would be more intense the closer a planet orbits to its parent star.

Until recently, astrobiologists believed that planets orbiting near red dwarf stars would be unlikely to harbour life. Planetary scientists had thought that the dark side of such a world would be so cold that all the gases in the

Artist's impression of the proposed infra-red interferometer, part of NASA's Terrestrial Planet Finder mission, which will be launched by 2020. The beams show incoming infra-red from distant planets.

atmosphere would condense onto it. This 'atmospheric collapse' would leave the whole planet airless, and therefore lifeless. As the atmospheric gases move from the light side to the dark side, they do carry heat with them, warming up the dark side to a certain extent. But theory suggests that this would only prevent atmospheric collapse if the atmosphere was very dense. The atmosphere would have to be so dense, scientists supposed, that not enough of the star's light would reach the planet's surface for life to survive. That view changed in the 1990s, when scientists used detailed climate prediction models, like those used to monitor and forecast weather and climate on Earth. As long as the planet has deep oceans, warmed from below by magma under the crust, liquid water can flow back to the light side, where it will eventually evaporate and travel back to the dark side. This water cycle makes it possible for an atmosphere and liquid water, and therefore life, to exist.

The red dwarf is not the only type of star that has recently been re-evaluated in terms of its potential for habitable planets. Many stars form with a companion,

One possible design for the European Space Agency's Darwin project, which also aims to find signs of life on planets around distant stars. It is due for launch around 2015.

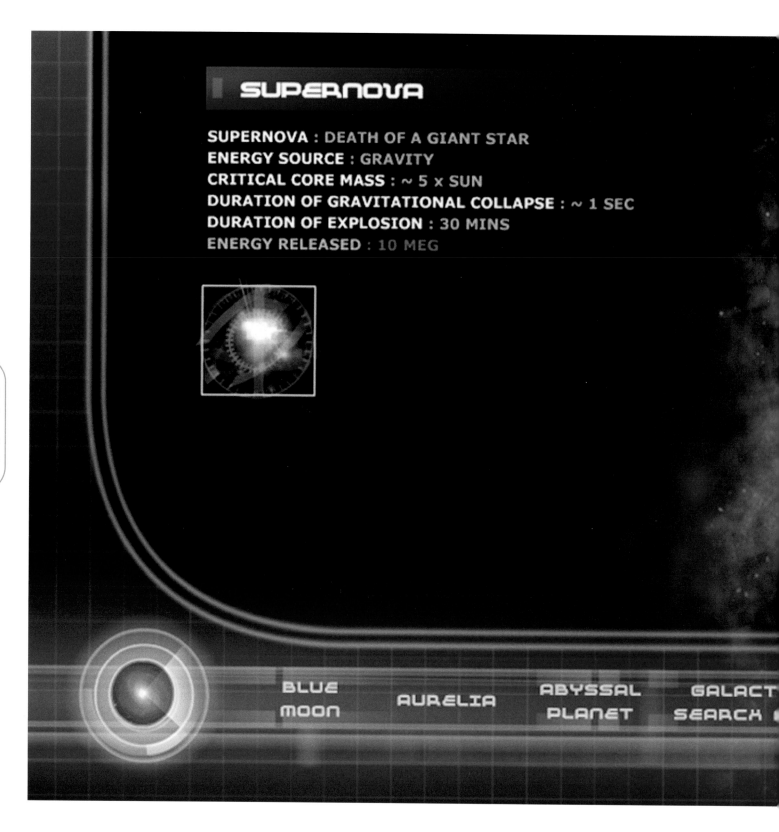

SUPERNOVA

SUPERNOVA : DEATH OF A GIANT STAR
ENERGY SOURCE : GRAVITY
CRITICAL CORE MASS : ~ 5 x SUN
DURATION OF GRAVITATIONAL COLLAPSE : ~ 1 SEC
DURATION OF EXPLOSION : 30 MINS
ENERGY RELEASED : 10 MEG

BLUE MOON   AURELIA   ABYSSAL PLANET   GALACT SEARCH

▲ All planets are made of just 100 or so chemical elements, produced by nuclear reactions inside stars. When a massive star explodes, as a supernova, the elements are spread across space, and are incorporated into new solar systems.

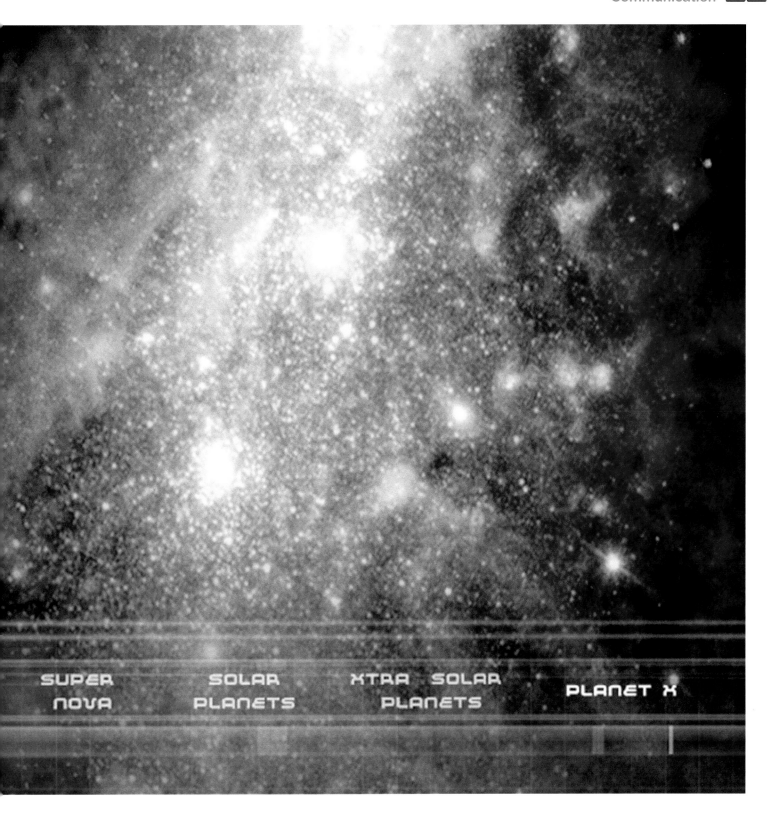

SUPER
NOVA

SOLAR
PLANETS

XTRA SOLAR
PLANETS

PLANET X

creating a binary system. The gravitational influence around a pair of co-orbiting stars is complex, and until recently astronomers believed that it would be impossible for planets to have stable orbits around a binary system. They believed planets would be hurled out into interstellar space. But new calculations have suggested that there are several configurations in which a planet could orbit for thousands of millions of years, long enough for life to develop.

When British company, Big Wave, commissioned ten scientists to create two imagined extra-solar worlds, the scientists chose to study what might happen around a red dwarf and around binary stars. The worlds they brought to life are called Aurelia and Blue Moon.

Aurelia – a World of Extremes

Planet Aurelia orbits close to a red dwarf star, much closer to its sun than the Earth to ours. From the planet's surface, Aurelia's reddish sun appears slightly larger than the Sun does in our sky, and never moves. Locked in a gravitational dance with its parent star, the planet has no seasons, and two very different sides. The dark side is very cold, trapped in a constant, deep winter. It is unlikely that life could survive there. By contrast, the centre of the light side is heated so much that a huge, slow-moving storm rotates constantly, resulting in hurricane-force winds and torrential rain.

The zones around the edge of Aurelia's dark hemisphere are mild. The planet has continents, with arid desert regions, as well as oceans and lagoons of liquid water. Living things thrive in the temperate zones. While many of the geological and climatic features of Aurelia are similar to those on Earth, Aurelian living things are very different from terrestrial ones.

Evolutionary biologists formed part of the team designing the planet. Applying the wealth of experience they have gained from studying life on Earth, what did they say about the kind of living things that might have adapted to life on this bizarre world?

As on Earth, life on Aurelia exists as a complex, inter-connected web of carbon-based organisms. Every living thing in that web requires energy. On Earth, the ultimate source of energy for nearly all life is the Sun: plants and some other organisms use energy from sunlight to make food, through the process of photosynthesis. On Aurelia, huge organisms called stinger fans harness the energy from the planet's red dwarf sun. Forests of stinger fans cover huge swathes of the planet's continents. These giants are more like animals than plants, but they share features of both.

Planet Aurelia

Red dwarfs were originally thought unable to support habitable planets. Cutting-edge research has turned this theory on its head. Red dwarfs do have planetary systems, but these planets pay a price: their rotation is slowed to a stop by the intense gravitational pull of their sun, half of the planet in perpetual darkness, the other in permanent light.

Aurelia is such a planet. On its dark side, a vast frozen waste; on its light side oceans of water, a temperate climate and land. In Aurelia's 'Twilight Zone', the sun's rays will start to dim and temperatures plummet. Creatures will have to compete for light, moving to high ground to capture the sun's rays. The sort of life that could survive on Aurelia is likely to be mind-blowing.

Their large fans are angled so that they can make the most of the sunlight, just like large plant leaves do on Earth. But these leaves are also jointed at the bottom, like animals' limbs. When they sense one of the red dwarf's violent ultraviolet flares beginning, muscles tug on them to close them up. Stinger fans have hearts, which pump water and the products of Aurelian photosynthesis around their circulatory system. At its base, a stinger fan has poison-releasing cells, which make them less attractive to predators – a feature that has evolved in several plants and animals on Earth.

Each stinger fan is actually made up of five individuals huddled together. The five leafy fans at the top of the organism form a circular structure six metres in diameter. A fully grown stinger fan is a giant, standing eight metres tall. These strange plant–animals live in wet, marshy areas on flood plains or next to ponds. They can move around slowly, thanks to slimy, muscular tentacles at their base. Just as plants are eaten by animals on Earth, Aurelian stinger fans are food for several different creatures. Whenever a stinger fan is felled, neighbouring stinger fans use their tentacles to jostle for position to make the most of the available sunlight. New stinger fans are created by asexual reproduction, new individuals emerging as buds on the trunks.

One creature that kills stinger fans and makes a meal of them is the mudpod. The skin of this six-legged amphibian incorporates the fans' venom thus making them immune to the fans' poison. The mudpod not only uses stinger fans as a source of food, it also dams rivers with them, to form ponds which it needs to keep its skin moist. Mudpods are Aurelia's landscape gardeners. They busy themselves, pushing and pulling soil around the edges of ponds to create new land, using their two front legs while remaining stable on

The view over one of Aurelia's stinger-fan forests.

the other four. The new land provides more space for stinger fans, the mudpods' staple food.

Mudpods have eyes, which are on stalks that protrude above the surface of the water when they are swimming in their ponds. The mudpods retract their eyes into the stalks when they are bulldozing Aurelian mud. Eyes are one of the features that have evolved many times here on Earth. The fact that they bring significant advantages gives evolutionary biologists reason to believe that they will evolve several times on other planets, too. Mudpods communicate with one another vocally, making shrieking sounds, and also visually, using brightly coloured fins on their backs which can stand erect as a sign of danger. They make burrows in the wet ground around their ponds. They use their burrows to hide away from two threats: ultraviolet flares from the sun and their greatest predator, the gulphog.

The bipedal gulphog is agile and voracious, and stands nearly five metres tall. A fully grown gulphog has a mass of half a tonne. It has keen eyesight, with camera-like eyes similar to our own. Its teeth provide its other main sense, picking up vibrations, especially those produced when mudpods fell stinger fans. They will travel long distances to find the vulnerable mudpod, which they can swallow in one dramatic gulp, thanks to their jaws, which can dislocate like a snake's. One ingenious invention of evolution by natural selection in this strange-looking species is a third 'eye' that is sensitive to ultraviolet. The special eye is located on top of the head, and quickly alerts the gulphog to the onset of a life-threatening flare. Instinctively, gulphogs run for cover, often hiding under a fallen stinger fan. Their gait as they run is reminiscent of any of the large, flightless birds on Earth. At the end of each of their long legs is a two-toed foot,

which is fleshy so that it spreads out, preventing the animal from sinking into the soft ground of the swamps in the stinger-fan forests.

The final Aurelian life form designed by the biologists is strangest of all. Hysteria is a tiny organism, with a diameter of a little less than a millimetre. It lives in the stagnant ponds created by mudpods. Hysteria has a spinning tail – similar to the flagellum found on many terrestrial single-celled aquatic organisms – which propels it through the water. It simply moves towards or away from chemicals dissolved in the ponds.

Normally a solitary organism, taking in nourishment from small microbes in the water, hysteria exhibits a bizarre collective behaviour when there is a lack of available food. This behaviour was based on a terrestrial organism called slime mould, which does a very similar thing. A colony of up to a million individual hysteria spontaneously swarms together, attracted by a chemical messenger they all release. The swarm acts as a single super-organism, which moves quickly through the water, and can even move a little way on to land. A toxin released by the hysteria super-organism can paralyse large prey – even a gulphog. Once their target is immobilised, the hysteria organism moves into the body, digesting it from the inside out, and reproducing rapidly.

Unbelievable as hysteria's behaviour might seem, something similar (though not so spectacular) exists on Earth. There are several species of aquatic single-celled organisms that form colonies of cells for a collective advantage. For example, a species called volvox forms colonies of up to 50,000 individual cells, all identical. Volvox colonies are in the form of hollow spheres. Somehow, the flagella of the

Profile: STINGER FAN

Stinger fans are Aurelia's unique animal-plants. They stand 8-m tall, and have an enormous 6-m diameter fan. These fans are permanently angled towards the red dwarf in the sky. The fan can be moved using tendons and muscles. During a UV flare the fan closes to protect itself from damage. The fan contains photosynthesising apparatus, which converts the red dwarf's rays to chemical energy. The products of photosynthesis are transported around the fan by an advanced vascular system. Half-way up the trunk are 5 hearts to pump the muscles, tendons and cartilage. When a fan is felled by a mudpod the remaining fans jostle for the rays. The base of the fan has muscular tentacles which can shorten and thicken to pull the fan. The base is protected from hysteria (see pp.90–91) by poison, however mudpods are immune to the stings and prey on the fans, using them to build dams.

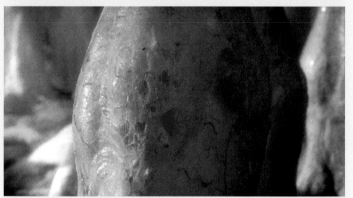

Chemistry	Carbon-based.
Life strategy	Alien animal-plant.
Trunk height	8 m/26 ft.
Fan opening	Muscles, tendons, cartilage.
Locomotion	Slow movement with tentacles. Muscles, tendons, cartilage at the base of the fan. 10 per fan. Single hook between each pair. Tentacle shortens and thickens in a slug-like way to pull the fan. Base secretes mucus to lessen friction. Underside of tentacle uses ridges and scaly skin, like a snake, to grip the soil.
Primary senses	UV detection.
Predators	Mudpod; gulphog.
Habitat	Wet environment of flood plains and mudpod ponds.
Energy production	Photosynthesis.
Defence	Stinging cells. Fan closes during UV flare.
Reproduction	Asexual. Young grow on trunk of adult.

Energy source	Solar radiation.
Energy conversion	Photosynthesis.
Products	Sugars.
Sugar distribution	Vascular system with 5 hearts.

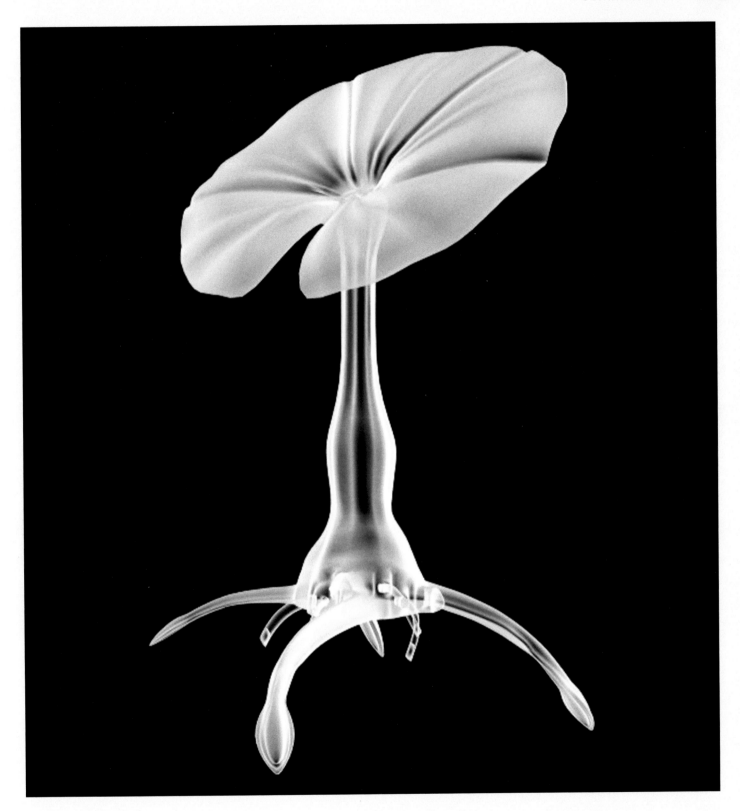

Mudpods are the keystone species of Aurelia. With the notched bony plates in their mouth, they fell stinger fans and use them to construct dams, in which they live. The dams create still freshwater pools, providing a habitat for a considerable number of species including hysteria. Mudpods are amphibious, so need this wet environment. They are immune to the stings from the fans, and actually incorporate the poison into their own flesh when they consume them.

They have vocal communication as well as brightly coloured 'flags' which can be drawn up and used for signalling danger. Their eyes protrude on stalks, allowing them to remain underwater but still able to see above it. During the UV flares they hide in their burrows. These also provide safety from gulphogs, although they can be rooted out.

They are strong swimmers using their powerful tail. On land they move in a similar way to a crocodile, using 3 pairs of legs.

Chemistry	Carbon-based.
Life strategy	Keystone species. Builds dams for stinger fans to grow on and then harvests them.
Body plan	Bilateral symmetry.
Size	0.9 m/3 ft.
Mass	10 kg/22 lbs.
Locomotion	Hexapedal walking. Swimming with tail.
Number of legs	3 pairs attached to the body in a lizard-like manner, low-slung at the side (not under the body).
Senses	Vision, sound and touch.
Favoured prey	Stinger fan.
Predators	Gulphog.

Defence	Poisonous spines. Stinger-fan venom in flesh.
Habitat	Burrows around pools in forest. Needs wet environment as is amphibious so loses a lot of water through skin.
Communication	Vocal. Has spine on the posterior dorsal surface. When erect draws up a brightly coloured flag used for signalling.
Legs	Three toes with thick toenail claws. Webbing between the toes for swimming. Front left foot has serrated claw for burrowing.
Tail	Long tail, similar to a crocodile for swimming. Two barbs on the dorsal surface.

The gulphog is a bipedal predator, standing almost 4.5 m tall. Its legs are powerful and muscular, with an extra joint, allowing complex locomotive control and top speeds similar to a galloping racehorse. The gulphog's feet have two toes, arranged in a V-shape that are soft and spread out under pressure, helping the creature to move about in the swamps. The toes have a horny serrated toenail that is used to dig for prey. Gulphogs use their teeth to sense for prey, detecting vibrations through the ground, just as elephants can detect storms and calls from other elephants through special cells in their feet. The prominent incisors are very sensitive to vibrations, especially those caused by the felling of trees by mudpods. Once tracked down the teeth are used to root out and consume them.

Their jaws are hinged to allow backwards and forwards movement. It can also dislocate like a snake to swallow prey whole. The gulphog has high-quality vision, with two camera eyes allowing depth perception and sharpness to approximately 1 km. The gulphog has an early-warning system, which allows it to find shelter as soon as possible. Gulphogs are social animals living in small clans of 10–15 animals. They communicate vocally. Their excellent vision and medium intelligence means that they have to sleep.

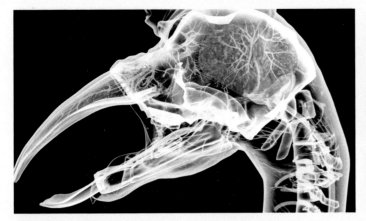

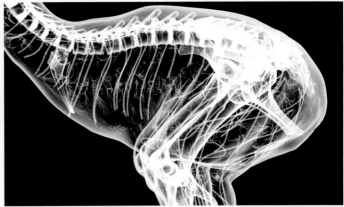

Chemistry	Carbon-based.
Life strategy	Predator.
Body plan	Bilaterally symmetrical
Mass	500 kg/1100 lbs.
Height	4.5 m/14.75 ft.
Locomotion	Bipedal.
Primary senses	Vision; vibration.
	Sensitive teeth. Sets of two pronounced upper and lower incisors. Very sensitive to vibrations, especially the felling of trees by mudpods. Used to root out and consume mudpods. Sensitive whiskers.
Eyes	2 camera eyes. Shade from UV above with little hoods. 3rd eye used to detect UV radiation.

Favoured prey	Mudpod; stinger fan.
Predators	Hysteria.
Habitat	Stinger-fan forests.
Communication	Vocal.
Intelligence	Similar to a baboon.
Locomotion	Bipedal.
Leg design	Two powerful long legs for fast movement and wading through water.
Maximum speed	61.2 km/h/38 mph.

Profile: HYSTERIA

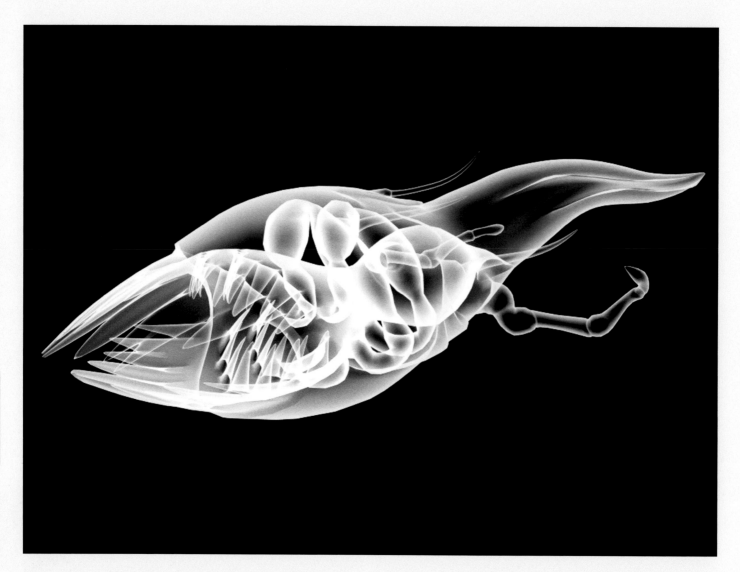

Hysteria is a multi-celled organism based on slime moulds and the dinoflagellate *Pfiesteria*. It lives in the still fresh-water ponds created by the mudpods. Under optimum conditions of plentiful food it remains as a single organism, using its whip-like tail for locomotion. However under food stress a chemical attractant is released from one end of the animal. This attracts the front end of another hysteria individual and leads to chains of individuals coming together.

This swarm may reach over a million individuals that now act as one super organism. Once it senses its prey the hysteria uses a paralytic poison, which prevents the animal from escaping. The hysteria floods over its body and begins to digest. Once the organism has fed, it produces a giant bud and releases spores into the atmosphere.

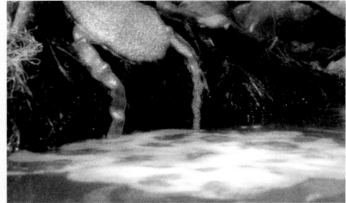
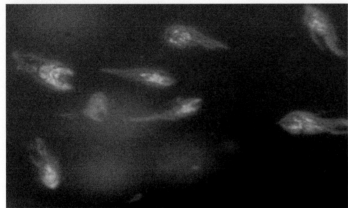
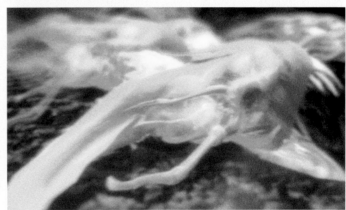

Chemistry	Carbon-based.
Life strategy	Predator.
Body plan	Multicellular and bilateral. Swarms under stress.
Cell size	0.5 mm/0.02 in.
Swarm size	> 1 million.
Locomotion	Active swimmer.
Favoured prey	Mudpod; gulphog.
Predators	None.
Habitat	Ponds in stinger-fan forests.
Communication	Chemical.
Toxin	Paralytic poison.

▲ Blue Moon, a jewel in orbit around two stars. The dense, oxygen-rich atmosphere gives the planet a bright blue glow where it is illuminated by its two suns.

individual cells are co-ordinated in order to propel the sphere much faster than any individual cell can move. On Earth, the association of single-celled organisms into groups was the beginning of the move to multicellular organisms; perhaps the same was true in the evolution of Aurelia.

Moving out from the forest community of stinger fans, mudpods, gulphogs and hysteria, and drawing away from planet Aurelia, we travel through interstellar space to the second of the imagined alien worlds. The living things we find there are just as bizarre as those we have seen on Aurelia.

Once on a Blue Moon

For the second of their hypothetical worlds, the team of scientists looked closely at something a little different. Blue Moon is not a planet: instead, as its name suggests, it is

Blue Moon

Two binary stars, each identical to our own Sun.... Orbiting them is a vast gas giant, 1,000 times bigger than Earth. Its Saturn-like rings are the remains of a moon that drifted too close and was torn apart by its intense gravitational pull. It is an uninhabitable world. But one moon orbits it at a safe distance, a strange world. Blue Moon glows like an oasis in the vast emptiness of space.

Enveloped by deep oceans, continents and an atmosphere it is a prime target for life. With three times the air density of Earth and high oxygen levels, Blue Moon is perfect for flying, gliding and floating. Life forms here will be unlike anything we know.

a satellite. This Earth-sized object orbits a massive gas giant planet many times the size of Jupiter; the gas giant itself orbits around both stars of a binary star system. Jupiter has 14 times the diameter of Earth, and 300 times the mass, so this alien gas giant is truly gargantuan. Most of the extra-solar planets already discovered are as big as Jupiter or larger. This does not necessarily mean that they are more common in the galaxy, simply that they are massive enough to have been discovered in planet searches that have already taken place. Gas giants themselves are unlikely to provide favourable conditions for life, but their moons might do. This is why space scientists are studying Europa and Titan, the moons of Jupiter and Saturn.

Blue Moon is much larger than any moon in our Solar System: it has almost exactly the same diameter as Earth and, like Earth, it is a rocky object with oceans of water and large continents. It is the dense atmosphere around the Blue Moon that is one of the main driving forces of evolution here, producing a rich diversity of plants and animals. Oxygen makes up 30 per cent of Blue Moon's atmosphere, compared to a little over 21 per cent on Earth. The availability of oxygen means that animals can be more energetic. Carbon dioxide is more concentrated in the atmosphere, too, and is 30 times more common than on Earth. The availability of carbon dioxide means that plant life grows in abundance.

On Earth, photosynthesising plants use the energy of sunlight to combine carbon dioxide and water, forming sugars. Plants are largely made of cellulose, which is formed from sugar molecules joined together. Vegetation on Blue Moon works in the same way, but it looks quite different from what

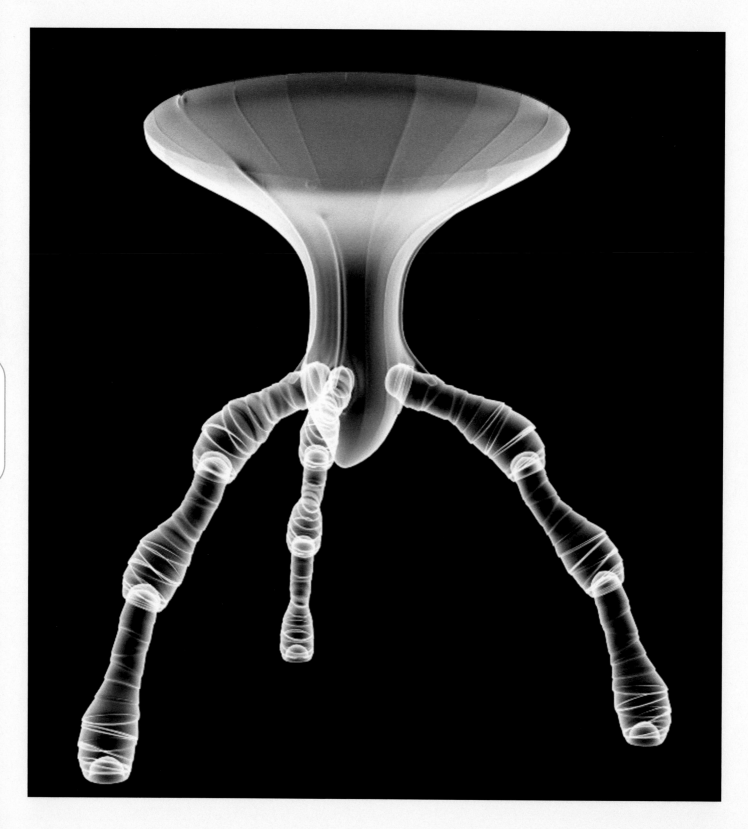

Pagoda trees form a giant cloned forest, which covers huge swathes of Blue Moon. With an incredibly strong network of branches that interlock they are able to grow to over 1,000 metres, more than seven times taller than the tallest tree on Earth, the sequoia. They collect water in their canopy with huge reservoir-type leaves that can contain over 2 m^3 of water. This is absorbed by rootlets in the leaves and an overflow into the branches. The main branches have smaller branches, which exit downwards. Water flows through the plant by passive water transport. The hollow tubes and trunk are ideal habitats for the colonies of stalkers. The leaves are also capable of photosynthesis.

Chemistry	Carbon-based.
Life strategy	Alien plant form.
Maximum height	1000 m/3280.8 ft.
Maximum height for tree on Earth	*Sequoia sempervirens*: 122–130 m/400–425 ft.
Energy source	Solar radiation.
Energy conversion process	Photosynthesis.
Light absorbed	Mainly infra-red, green and yellow. Blue light reflected.
Water source	Collects rainwater in the canopy.

Reproduction	Sexual and asexual. Fruiting bodies develop at the base of the leaves every 3 Blue Moon years.
Speed of growth	Slow.
Leaf diameter	2 m/6.5 ft. Water collected in large jug-like reservoir leaves. Up to 2 m across.
Leaf volume	> 2.1 m^3/74 ft^3
Water absorption	Through water-absorbing rootlets and overflow inlets in the leaves. Water flows down inside the branches to the rest of the tree right down into the main stems.

we see on Earth. Giant forests up to a kilometre high dominate the landscape. The tallest and most prominent species is the pagoda tree. On Earth, there is a limit to how high water can be taken up from the roots to the top of a tree. There is no such limit to the height of pagoda trees: they have evolved large basins in their canopies, called sky ponds, which collect rainwater. Beneath the canopy, the trunks of several individual trees are interconnected, giving them the strength they need to support the weight of the sky ponds, and holding them fast in strong winds.

Scientists have calculated that the atmosphere of Blue Moon would provide three times as much life as the Earth's atmosphere. The atmosphere of Blue Moon differs from Earth's atmosphere in another respect: it is far more dense. Planetary scientists believe that dense atmospheres may be common among extra-solar planets and moons. Titan's atmosphere is more dense than Earth's, for example. The atmosphere on Blue Moon is so dense that animals and plants that take to the air can do so much more easily than they could do on Earth. One amazing creature that has evolved to take advantage of this is the kite. Its body is large and flat, with an aerofoil cross-section like an aeroplane wing.

As wind blows over the kite's body, it produces an upward force, called lift, which stops it from falling downwards. Even though a kite can grow to more than ten metres across, it can stay up in the air even in a light wind. At any given wind speed, the dense air on Blue Moon produces much more lift than the air on Earth. In order to anchor itself, the kite has a tether at the front of its body, which it wraps around the pagoda canopy. The tether splits into three at the end, and each of the three 'arms' moves independently, enabling the kite to pull itself across the canopy. The kite has tentacles, which dangle in the sky ponds at the top of pagoda trees, fishing for this strange creature's prey: helibug larvae. The adult helibug is a strange creature that has threefold symmetry, unlike the two-fold (bilateral) symmetry common in terrestrial creatures. It has three helicopter-style wings, three eyes and three legs. It hovers above the surface of the sky ponds, with its head hanging downwards between its legs, looking for its prey. From time to time, it dives into the water, grabbing smaller organisms with its three legs.

The most majestic of all flying creatures on Blue Moon is the huge skywhale. It is completely blind, relying upon sonar to find its way around. This animal rises on thermals of dense air, then glides down again, although it flaps its vast

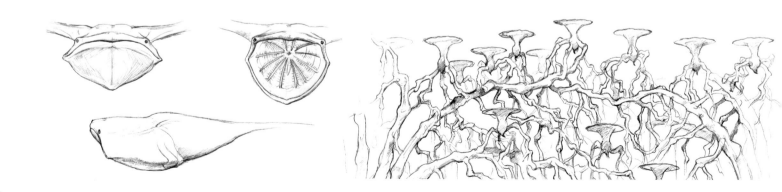

▲ The skywhale's huge mouth is built to scoop prey like a basking shark; pagoda trees have an enormous network of branches, which create a giant canopy.

wings, too. These thermals are created because some parts of the land heat up more than others. There are also updraughts of air where the wind blows into cliff faces. The skywhale looks like a whale, but with wing-like flippers that have adapted to gliding in dense air rather than in the oceans. The evolutionary biologists on the design team reasoned that life on Blue Moon would have begun in the oceans, just as on Earth. But they supposed that evolution might take some life forms straight from the ocean to the air, since the atmosphere is so dense. On Earth, animals first left the oceans about 350 million years ago, but it was another 150 million years before large reptiles took to the air (although insects could fly by 300 million years ago).

The skywhale is ten times heavier than any animal that has ever flown on Earth. Our planet's largest flyer was the pterosaur – a huge dinosaur that had evolved a membrane between each of its front legs and its body. The pterosaur had enough energy to manoeuvre such a large body because Earth's atmosphere contained much more oxygen than it does today. On Blue Moon, there is an even greater concentration of oxygen, and it is at high pressure thanks to the dense atmosphere. So, the biologists reasoned, there would indeed be giants in the air on Blue Moon.

As the skywhales glide, their open mouths trawl the air for tiny plant-like organisms – the equivalent of plankton that many whales trawl for in Earth's oceans. Blue Moon's carbon dioxide-rich atmosphere makes it possible for these tiny green cells to exist in abundance, and they are buoyed up in the dense air, and carried upwards by the same thermals that lift the skywhales. As the skywhales drift through air tinged green with the floating 'plankton', they also take in helibugs that stray too high above the sky ponds below. The life of a skywhale is slow and serene – but it is not without danger.

Just as the high concentration of oxygen in Blue Moon's atmosphere means that a huge beast like the skywhale has enough energy to control its flight, it also supercharges its only adversary: the stalker. Living in colonies of several hundred individuals inside the trunks of pagoda trees, stalkers are similar to social insects such as wasps and hornets on Earth. Three different roles have evolved in one species. There is a queen, which is responsible for producing large numbers of offspring to maintain the colony; there are workers, which swarm into the sky to kill skywhales; finally there are scouts, whose job it is to locate skywhales that are flying low in the air. The scouts have three eyes, spaced evenly around their head, so that they have 360-degree

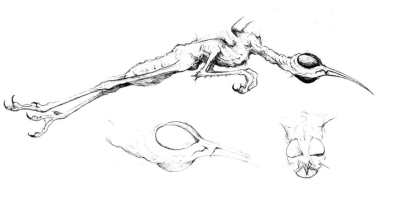
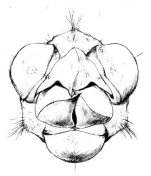

The stalker scout's strong eyesight and sense of smell enables it to detect prey, such as the skywhale; the stalker has 3 large compound eyes and 3 parts to its beak from which 3 tongues emerge to scent prey; the helibug's body is trilaterally symmetrical, with 3 legs and 3 wings on the dorsal surface.

Profile: KITES

Chemistry	Carbon-based.
Life strategy	Predator.
Body plan	Bilateral symmetry.
Mass	0.2–5 kg/0.44–11 lbs.
Locomotion	Gliding. Moves through canopy with its tether.
Wingspan	1–5 m/3.28–16.41 ft.
Primary senses	Touch.
Predators	None. Has toxic flesh.

Habitat	Pagoda canopy.
Prey capture	Tentacles capture food and haul it into multiple mouths all leading to a central stomach.
Aerodynamics	Wings can be drawn in to adjust to stronger wind speeds, or let out to stay airborne in light winds.
Tether and feet	Muscular feet at the end of the tether can move through the canopy by latching onto new branches.

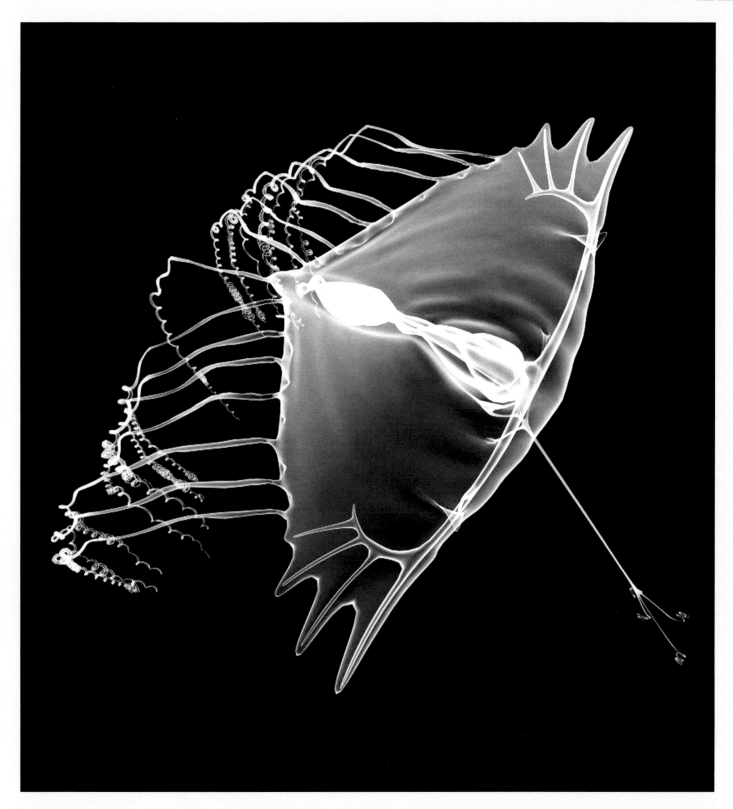

Profiles: GHOST TRAP AND HELIBUG

Life form	Ghost trap.	Prey capture	Drapes a sticky veil across the canopy to snare flying prey. Tentacles draw the prey into the mouth and acidic bath.
Chemistry	Carbon-based.		
Life strategy	Predator.	Digestion	Sharp teeth attached to a muscular tube. Downward pointing to ensure no escape. Retractable teeth roll out of the mouth and claw the prey into the stomach. Mouth tube rolls out of opening; mouth closes to digest food. Body is a sac full of digestive juices. Once inside the food is chopped up and crushed, by waves of contractions.
Body plan	Radial symmetry.		
Locomotion	Slow-moving tentacles.		
Primary senses	Touch.		
Favoured prey	Caped stalker.		
Predators	None.		
Habitat	Pagoda forest.		

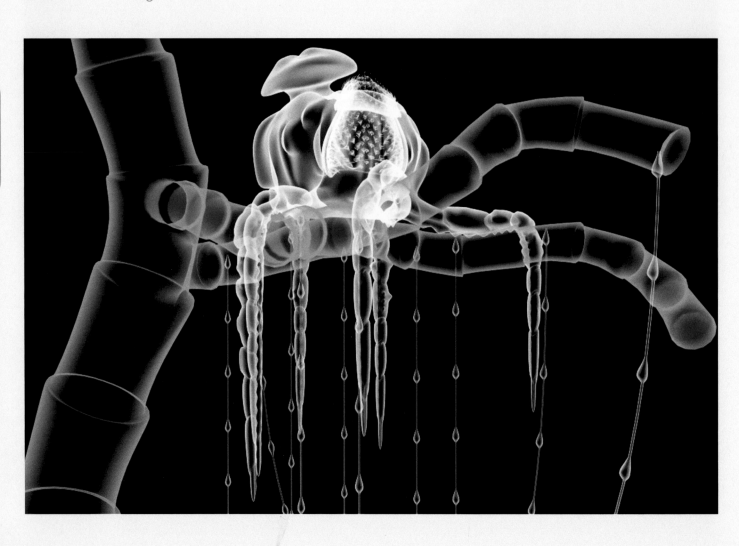

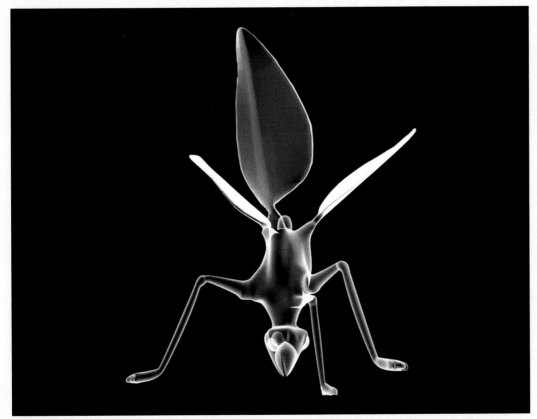

Life form	Helibug.	**Primary senses**	Vision.
Chemistry	Carbon-based.	**Eye design**	3 compound eyes. 360° peripheral vision.
Life strategy	Predator. Hovers above pagoda leaves and dives in to capture alien larvae.	**Predators**	Skywhale.
Body plan	Trilateral symmetry. 3 wings, 3 legs and 3 eyes.	**Habitat**	Pagoda forest.
Mass	0.003–0.2 kg/0.007–0.44 lbs.	**Communication**	Pheromones and bioluminescence at night.
Locomotion	Hovering flight and can swim in water to capture prey.		
Wing design	Shaped like hummingbird wings. Plane of beating tilts forward for forward flight. Each wing beats $\frac{1}{3}$ cycle out of phase with its neighbours.		

Skywhales are giant aerial grazers that scoop up small aerial plants and animals with their enormous mouths, and filter them. The air leaves through gill-like slits on the side of the head. Weighing in at 600 kg, but capable of reaching 800 kg they are the same weight as a small killer whale, and are 7.5–10 times heavier than anything that has ever flown on Earth. Their 5-metre-long (10-m wingspan) wings are designed for strength and to be lightweight. They are a strengthened honeycomb structure of gas-filled cells. If one of these cells bursts, the wing will still remain rigid. The wing is incredibly aerodynamic, with an elliptical cross-section to improve its ease of movement through the air.

The leading edge of the wing is supported by a thin rod of cartilage. There is a main ball-and-socket joint allowing all-round movement and the wings are very muscular. Although they can fly quite rapidly (22 mph/36 km/h) this alone is not fast enough to escape from their main predator, the caped stalker. They have to use thermals to soar higher and higher out of the stalkers' reach.

Skywhales are social and nurse juveniles, who glide along in the slip-stream of their parents, and latch on to feed. They use echolocation, sonar and acute hearing to explore their environment as they are completely blind.

Chemistry	Carbon-based.	Prey capture	Filters plankton with its giant mouth.
Life strategy	Grazes on aerial plankton.	Communication	Sound emission from balloons on head.
Body plan	Bilateral symmetry.	Wing design	Similar to the fuselage of an aircraft, it is strengthened by a honeycomb structure of gas-filled cells. If several cells burst, it can still fly.
Size	5.25 m/17.2 ft (body length).		
Wingspan	10 m/33 ft.		
Mass	600 kg/1320 lbs.	Wingspan	10 m/33 ft.
Locomotion	Gliding and flying on thermals.	Tail	Fanned shape for stability and propulsion.
Primary senses	Hearing. Uses echolocation like a dolphin or bat to navigate.	Maximum speed	1 beat every 3 seconds.
Favoured prey	Aerial plankton and helibugs.	Maximum wing beats	0.8–0.67 cycles per second.
Predators	Caped stalkers.	Largest flying animal in Earth's history	*Quetzalcoatus* (pterosaur, mass: 80 kg/175 lb; wingspan: 11–12 m/36–40 ft.).
Habitat	Gliding on thermals high above the pagoda forest.		

Profile: STALKER

The stalker is a voracious social predator. It lives in colonies in the hollow trunks of pagoda trees. The colonies may comprise up to 300 stalkers, of which only 5 per cent are scouts. Scouts have an excellent sense of smell and 3 large compound eyes, which can detect fast movement, very useful for high-speed flight (up to 50 km/h/30 mph) through the pagoda canopy.

They have 360-degree vision, which can also detect polarised light reflecting from the bodies of their main prey source, the skywhale. Each eye has 150,000 facets, which are very small (0.1 mm), but do not provide good resolution. When hunting, groups of scouts set out from the nest and fly through and above the canopy looking for prey.

Once detected the scouts scent-mark the skywhales with a potent chemical. They then return to the colony. The workers will then flock from the colony, following the scent trail with their excellent sense of smell. Their 3 tongues pick up the smallest particles, and different concentrations allow the direction of the prey to be determined. Arriving in a huge flock the stalkers attack the skywhale viciously, biting it with their razor-sharp beaks, removing chunks of flesh. They destroy the wings, preventing the creature from escaping and in shock it falls onto the canopy. It provides a feast for the stalkers, who bring food back to the colony for their queen. These animals are highly intelligent and socially complex animals.

Chemistry	Carbon-based.		Castes	Scout; worker; queen.
Life strategy	Social predator.		Flight	Rapid (max. speed 50 km/h/30 mph) and highly manoeuvrable.
Body plan	Trilateral symmetry.			
Mass	0.4 kg/0.9 lbs.		Wing design	Membrane wing stretched between elongated bones.
Locomotion	Flight.			
Primary senses	Vision, smell and taste (has 3 tongues).		Wingspan	1 m/3.28 ft.
			Maximum wing beats	3 cycles per second.
Favoured prey	Skywhale.			
Predators	Ghost traps.		Intelligence	High, based on collective intelligence of social insects.
Habitat	Pagoda forest. Burrows in giant tree holes.			
Origins	Marine. Evolved from creatures similar to a flying squid.		Eye design	3 large compound eyes. 360° peripheral vision.
Colony size	300 stalkers.		Legs	4, long with talon-like claws at the end.

Profile: BALLOON PLANT

Balloon plants are large photosynthetic bladders that float high in the sky. Their height is controlled by an adjustable, telescopic tether, which allows the bladder to rise and fall. They are filled with hydrogen gas, which is produced by symbiotic water-splitting bacteria. These bacteria are housed in special sacs with elastic walls. Water passes easily into the sacs ready for a new build-up of H_2 pressure. The outside wall of the bladder is photosynthetic, the innermost layer is formed of gas impermeable guanine crystals.

Chemistry	Carbon-based.
Life strategy	Alien plant form. Invasive species.
Body plan	Radial symmetry.
Maximum mass liftable	232.5 kg/512 lbs.
Balloon diameter	5 m/16.4 ft.
Maximum height	80 m /262.5 ft.
Height control	Adjustable telescopic tether.
Primary senses	Stretch. Pressure.
Habitat	Invades forest gaps cleared by fire.
Balloon lining	Outside photosynthetic layer, middle elastic layer and internal layer of guanine crystals to prevent loss of hydrogen. Pressure can be released from balloons in strong winds, so they can drop to the ground.
Balloon structure	Between mother balloon and juveniles, called candelabra. Stiffer sections of stems can be pulled together by osmosis, bringing balloons together in bad weather. Should a balloon become damaged the stiffened section could break at joint with tether and be quickly replaced by waiting buds.
Trunk	Long telescopic tether that allows bladder to rise and fall.
Roots	Large mangrove-like heavy base that is well anchored.
Reproduction	Balloons are released from the plant. They drift into fire, explode and release their seeds to colonise new gaps in the forest. New balloons are formed from buds at the base of the mother balloon.
Water supply	Drawn up through roots and stored in sacs at the base of the balloons. Dew also absorbed directly through the balloon.

vision. They are agile flyers, and when they locate a potential meal for the colony, they mark it with a potent scent, which guides the workers. The workers bite into the sky whales' flesh with powerful beaks and destroy the huge wings, so that the sky whale falls into the canopy. The stalkers devour their prey, taking chunks of flesh back to the colony's nest.

But there are dangers, even for the stalkers, amid the tangled trunks of the pagoda trees. The ghost trap is a plant-like organism with many tentacles. It sits in the pagoda branches, waiting for stalkers to pass nearby. The ghost trap also has a sticky 'veil' not unlike a spider's web. When the prey touches the veil, it sticks, and the tentacle lifts its prey into a large vat of corrosive digestive juices. A set of sharp teeth unrolls to ensure that the stalker cannot escape. Once inside the ghost trap's primitive stomach, the stalker is dissolved by the acidic digestive juices, as the trap's stomach walls contract.

There is one phenomenon on this alien world that is a danger to most life forms: forest fires. With such a high concentration of oxygen in Blue Moon's atmosphere, and lightning storms commonplace, raging forest fires occur regularly. Fanned by winds, these fires destroy large areas of

pagoda forest, leaving behind a charred wasteland. But just as on Earth, life soon takes hold again in areas ravaged by fire. In terrestrial forests, small plants that normally struggle for existence, as they are overshadowed by spreading tree canopies above, thrive in the clearings left after forest fires. On Blue Moon, the same thing happens, but the plants are, of course, quite different from those on our planet. The main species of plant that takes over after destruction of the pagoda trees is the balloon plant. As its name suggests, the most obvious feature of a balloon plant is a large bag of gas. Even this strange phenomenon is based on similar examples on Earth. Kelp is an aquatic plant that has gas-filled bladders along its stems. These bladders give buoyancy to the plants, making them stand up straight in the water, reaching for the sunlight that penetrates the water.

The gas bag of a balloon plant on Blue Moon is filled with hydrogen gas, produced by bacteria inside the plant, which break down water molecules. There are many examples of organisms here on Earth which produce hydrogen gas in this way. The relationship between the balloon plant and the bacteria is beneficial to both species – what biologists call symbiosis. The plant provides the perfect environment for the bacteria, while the bacteria produce the gas that

▲ The mudpod has poisonous skin as well as a poisonous spine. The serrated claw is for cutting down stinger fans.

plays such an important role in the lives of the plants. The hydrogen-filled balloons buoy up the plant to catch more sunlight. The balloons are secured to the ground by long, telescopic tethers. As the pagoda trees begin to establish themselves again, the tethers lengthen, so that the plants float higher, ensuring their survival. Eventually, the gas bag is released, floating away to colonise elsewhere.

The plants and animals on Aurelia and Blue Moon may seem unbelievable. But all of them – and the relationships between them – have been designed to the very last detail by experts in evolutionary biology, biomechanics and planetary science. Organisms such as the ones described here really could exist. If you knew nothing of life on Earth, and someone described to you some of the things that live here, you would probably think they were unbelievable, too.

Evolution produces convergent features that have a good chance of arising on other planets, enabling evolutionary biologists to make educated guesses about alien life. High intelligence might even be one of those features. Perhaps a planet like Aurelia, with its very long-lived sun, could develop highly intelligent life. The evolutionary biologists in this study decided not to give any of their creatures high intelligence, as there is no way they could make educated

guesses as to what that might be like. If there are intelligent creatures on planets around red dwarf stars, they too might speculate about what life might exist on other planets. They might look at shorter-lived stars like the Sun, with planets as far away from their stars as the Earth is, and think that there would be no way that life could survive there, let alone be intelligent.

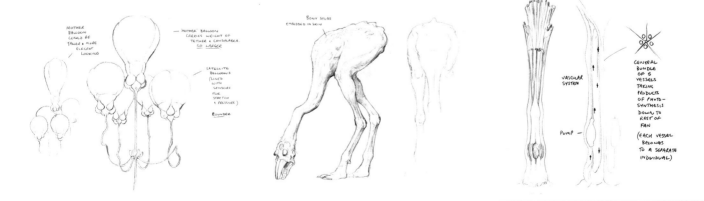

▲ The larger 'mother' balloon plant carries the weight of both the tether and the candelabra; the gulphog has fur on its back and bony studs embedded in the skin; a closed stinger fan, its central bundle of 5 vessels carries the products of photosynthesis down to the lower tentacles.

Gold–aluminium cover designed to protect a gold-plated record with sounds and pictures from Earth. One disc was carried by each of the Voyager space probes.

Alien Communication

It is tantalizing to think that alien 'plants' and 'animals' like the ones featured in the last chapter might really exist out there in another solar system. If any of the searches for Earth-like planets reveal a promising location for life, scientists would be very keen to pay a visit. It would be very exciting to be part of an expedition to discover such life forms. But, as we shall see, the practicalities of mounting such a mission make it almost impossible to manage it in the lifetime of anyone alive today.

Another, equally inspiring prospect is that some extra-terrestrials might be intelligent – even more intelligent than humans. If they are, they too will realise the physical limitations on expeditions to planets outside their own system. But that does not mean that we shall never know of their existence.

The Final Frontier

Space, as *Star Trek* correctly points out, is the ultimate undiscovered territory. But exploring the territory of space beyond the Solar System is daunting: the size of the Universe is mind-blowing. The nearest star, Proxima Centauri, is more than four light years distant. To gauge just how far this is, consider that light travels 300,000 kilometres in one second – equivalent to more than seven times around Earth. In the time light takes to reach us from the nearest star, it travels nearly 40 million million kilometres; that is the distance to Proxima Centauri. Of all the space probes sent into space from Earth, four are travelling fast enough to escape the Sun's gravitational influence. Voyager 1 is the most distant. Launched in 1977, Voyager 1 is on a course into deep, interstellar space. It is travelling at a speed of 62,000 kilometres per hour (17 kilometres per second). In 2020, after a journey of more than 40 years, Voyager 2 will have travelled over 20 thousand million kilometres. This tremendous distance is just 0.05 per cent of the distance to

The Science of Aliens

▲ The USS Enterprise, from the *Star Trek* television series and feature films. It is able to travel immense distances across space using a warp drive powered by dilithium crystals.

Proxima Centauri; if the space probe were travelling in the direction of Proxima Centauri, it would take nearly 80,000 years to reach its destination.

With conventional rocket technology, speeds twice that of Voyager 1 would be possible. But that would still take a spacecraft 40,000 years to reach even the very nearest star. Several innovative technologies have been suggested to increase the speed at which a spacecraft can travel. In the 1970s, a group of British scientists examined the possibilities of travelling to a nearby star. They suggested using nuclear fusion to create explosions which would propel a spacecraft up to speeds of more than ten per cent of the speed of light. One suggested fusion rocket would even collect hydrogen – the fuel for the most common form of fusion – in interstellar space along the way. Using fusion-powered rockets, a one-way voyage to Proxima Centauri would take nearly 50 years.

Another technology that could be used for interstellar travel is space sailing. Using the pressure exerted by sunlight, solar sails have already been tested successfully. Huge, extremely thin, reflective sails reflect the sunlight, and this pushes the sail in the opposite direction. Light exerts only a very slight pressure, but the cumulative effect over long periods means that very high speeds could be attained. In order to benefit from this light pressure at great distances from the Sun, one scheme suggests building very powerful solar-powered lasers in orbit near to the Sun, and sending the highly focussed beam in the direction of the spacecraft. Detailed studies suggest that, using space sailing, it would be possible for a spaceship to

The Science of Aliens

carry people to a nearby star. Sustaining human pioneers on such a mission provides another set of enormous technological difficulties, but it is an exciting idea. The journey time would be about the same as that for fusion-powered rockets. For longer journeys, humans could be in some kind of suspended animation – if that proves possible – or human embryos could be frozen and then grown in time for the arrival at the stellar destination. Schemes that involve several generations of human pioneers would raise serious ethical issues.

Suggestions for interstellar travel at still higher speeds – even faster than light – have been suggested. All of these are likely to be far beyond the technological capabilities of humans for hundreds or thousands of years, and most of them rely on theoretical physics that may be proved wrong

in years to come, or may be simply not workable in practice. One idea was put forward by Mexican physicist, Miguel Alcubierre, in 1994. Using Einstein's *General Theory of Relativity*, Alcubierre's proposal involves 'warping' space around a spaceship, so that space is contracted in front of the ship and expanded behind it. The occupants of the spaceship would effectively be propelled at faster than the speed of light in a 'space-time bubble', and could reach distant stars in an arbitrarily short amount of time. While the idea is in keeping with the best theories of physics today, it does rely on something called 'negative matter', which may not exist. Another approach to interstellar travel that is popular with science-fiction authors is the use of wormholes. These 'tunnels' from one point of space and time to another are predicted by the mathematics of

▲ The view from the cockpit of a hypothetical spacecraft travelling at eight-tenths the speed of light. At such extreme speeds, the stars ahead appear wrapped around the spacecraft, and their light is 'blue-shifted'.

General Relativity, but even if they exist, and it were possible to use one, or even build one, you would never know where in space-time you would emerge.

None of the very nearest stars is known to have planets. Most stars in our galaxy are tens, hundreds or thousands of light years away; the nearest major galaxy outside our own lies more than two million light years from us. However daunting the technological challenges of interstellar travel might be, the curiosity, ingenuity and pioneering spirit of humans is such that people will probably one day make epic journeys to other stars. But with the barriers that stand in our way, that may not happen for many thousands of years, if they ever do. Nevertheless we might still be able to discover extraterrestrial life in other planetary systems – if the inhabitants are intelligent and want to get in touch.

Tuning in

The idea that intelligent, extraterrestrial beings might want to communicate, or that we might be able to communicate with them, is not a new one. As long ago as the 1820s and 1830s, several people began to suggest schemes by which we could alert the inhabitants of the Moon and Mars to our existence. Ideas included planting forests in the shape of right-angled triangles, and building huge mirrors to reflect sunlight. In 1891, a prize of 100,000 francs was announced by French astronomer and author Camille Flammarion to anyone who could communicate with extraterrestrials and receive a response. With the invention of radio, new ideas were put forward. As early as 1901, Serbian–American scientist and inventor Nikola Tesla began designing an apparatus that he hoped would

send radio signals to Mars. In 1924, American astronomer David Todd suggested to the US Army and Navy that they stage a radio silence, so that he would be able to detect radio signals from Mars. The military granted him his wish, and allowed him to use their receiving equipment, but no signals were detected.

Astronomers began using radio telescopes to detect radio waves from space in the 1930s. These radio waves are not messages from intelligent aliens; instead, they are produced by stars and by interstellar gas. Radio astronomy quickly proved to be an invaluable tool in modern astronomy. Just like a radio or a satellite receiver, radio telescopes must be tuned in to a particular frequency. Early on, astronomers realised that it would be sensible to tune in to radio waves produced by hydrogen atoms. By far the most abundant element in the Universe, hydrogen emits radio waves at several different frequencies, one of which is in the radio part of the spectrum. This has a frequency of 1420 MHz (megahertz), that is, radio waves with a wavelength of 21 centimetres. In 1959 two physicists suggested that other intelligent civilisations, if they exist, would also be studying radio waves from space, and that they too would be tuning

▲ A hypothetical hyper-fast spacecraft, which uses 'negative energy' to warp space around itself. The technology, while out of reach for the foreseeable future, is permitted by the laws of physics as they are currently understood.

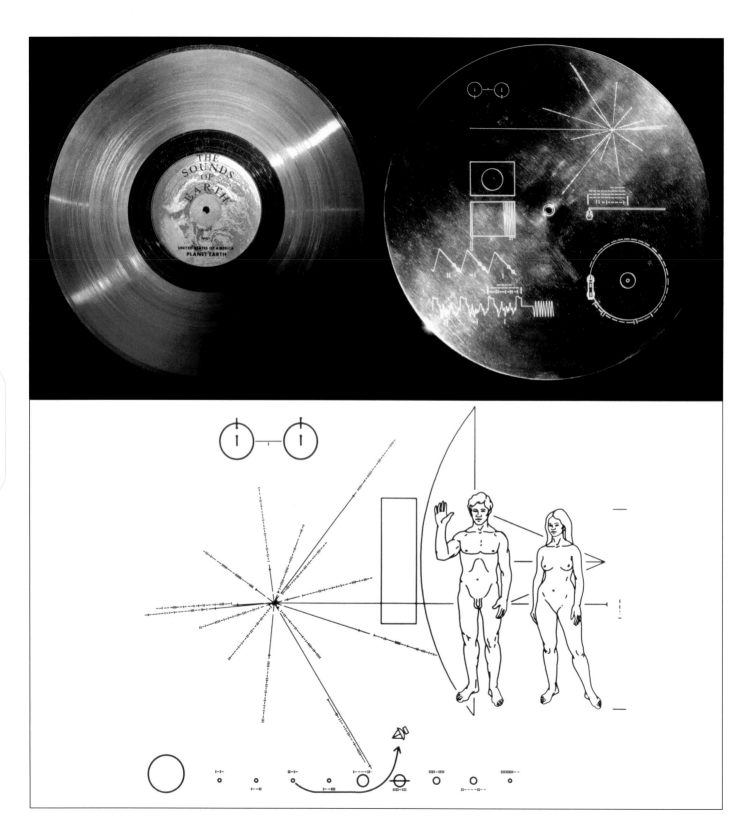

Top: the golden disc aboard the Voyager space probes, together with its protective and informative cover. Bottom: the image carried aboard the Pioneer 10 and 11 space probes, carrying information about humans and their planet.

Messages to Other Worlds

Ordinary radio and television signals have been leaking into space for more than 100 years. However, the power of these weak signals diminishes as they disperse in all directions through space, and the chance of any intelligent extra-terrestrial beings picking them up is virtually nil.

Even messages that humans have intentionally sent out into space have very little chance of ever being picked up. The 30-centimetre gold-plated copper disc carried by the two Voyager space probes contains 115 images, together with natural sounds, music and spoken greetings in 55 languages. The cover holds instructions on how to play the record, and information about the location of Earth. Attached to each disc are a cartridge and a needle.

Binary-coded information carried on the 23-centimetre-wide aluminium plaque attached to the Pioneer probes was designed to inform intelligent creatures about where and when the plaque was made, and by what kind of beings.

▲ The 305-metre-diameter Arecibo radio telescope is involved in the search for extraterrestrial intelligence, but it has also broadcast its own powerful signal into space. In 1974, it delivered a binary-coded message in the direction of a star cluster 25,000 light years away.

their radio telescopes in to the '21-centimetre hydrogen line'. Not only would they be listening, suggested the physicists, they might have gone one step further and decided to broadcast an intelligible signal. At about the same time, American astronomer Frank Drake came to the same conclusion.

Frank Drake is one of the main players in the search for extraterrestrial intelligence (SETI). In 1961 at a scientific conference in Green Bank, West Virginia, he formulated what has become known as the 'Green Bank Equation'. The equation provides a way of estimating how many intelligent civilisations might be capable of communicating via radio signals. The equation is formed from the following chain of logic: of all the stars formed in our galaxy, only a fraction will develop planets; only a fraction of those planets will be suitable for life; of those planets, life will only appear on some; on only a fraction of those will intelligent life develop; only a fraction of intelligent civilisations will develop radio astronomy. Finally, the equation takes account of the possibility that intelligent civilisations will not last forever: they may destroy themselves by warring, or they may be subject to catastrophic events such as the impact of a comet or meteorite.

Of all the terms in Drake's equation, only the first one – the rate of star formation – is known with any degree of accuracy. Since we know of only one species intelligent enough to engage in radio communication, no one can have any real idea of what fraction of intelligent civilisations might also do so, or even if any such civilisations exist. The point of the equation was to stimulate debate and focus thinking, not to attempt to work out how many civilisations might be ready to communicate. Nonetheless, various scientists have offered their own estimates of the numbers involved in the equation, and came up with a range of estimates for the number of civilisations that might be able to communicate. The range was between one and many thousands.

In the year before the Green Bank conference, Frank Drake conducted the first of many searches for radio signals from alien civilisations: Project Ozma. For a total of 150 hours, Drake pointed a large radio telescope in the direction of two stars similar to the Sun, and tuned in the telescope across a range of frequencies either side of 1420 MHz. He detected no unusual signals. Since Project Ozma, there have been around a hundred SETI searches, with increasing sophistication. Radio telescopes now scan millions of frequencies at a time, like listening to a million different radio stations at once. These searches carry out the equivalent of Project Ozma in a fraction of a second.

The most important and long-running SETI project is SERENDIP (the Search for Extraterrestrial Radio Emissions from Nearby Developed Intelligent Populations). Project SERENDIP is a long-term project that started in 1979. It rides piggyback on the world's largest radio telescope, at Arecibo in Puerto Rico. A receiver listens for radio emissions from whichever part of the sky the telescope happens to be pointing. Since the end of 1998, some of the data from Project SERENDIP have been part of a remarkable project called SETI@home, which involves more than five million internet users in 210 countries. The participants' computers gather SERENDIP data when they are connected to the internet, and analyse that data when their processors are

not busy running other programmes. This 'distributed processing' makes the system the most powerful computer in the world. The same computer programme also carries out analysis of data in other scientific fields, including cancer research and climate prediction.

The sky is huge, and there are millions of stars in our galaxy, so SETI will take a long time to complete. An intelligent civilisation would be unlikely to broadcast a signal in all directions, like a beacon. If they did, the signal would disperse, and would be extremely weak by the time it reached other stars. So those involved in SETI hope that alien civilisations will direct their broadcasts to many 'candidate' systems, including the planetary system around the star that we call the Sun. If broadcasts are being made, and heading in the right direction, they might be intermittent. In that case, not only must radio telescopes here on Earth be pointing towards exactly the right region of the sky and listening at the right frequency, they must be listening in at just the right time, too. The chances of detecting a genuine message from an extraterrestrial civilisation are very slim.

▲ Based on the Mayan calendar, this crop circle mysteriously appeared in a field near to the neolithic burial site at Wayland's Smithy, Oxfordshire.

▲ This 'Alien Face' crop circle appeared in a field in Hampshire in August 2002. The circular formation can be read as a spiral of binary digits, which spell out a message in English.

There have been several signals during the history of SETI which have aroused interest. Most are now known to have been false alarms. Perhaps the most famous was received by Irish astronomer Jocelyn Bell, who was investigating distant galaxies, but not engaged in SETI. While using a radio telescope in Cambridge, England, in 1967, Bell noticed a rapid and extremely regular radio pulse. It was like nothing anyone had seen before, and the regularity of the pulses made it seem unlikely that this was from a star or galaxy. As a joke, Bell wrote 'LGM' on the paper readout – signifying 'Little Green Men'. It turned out that Bell had discovered a type of dead star called a pulsar. In 1977 during an all-sky SETI search, American astronomer Jerry Ehman scrawled 'Wow!' on the printout of signals from a radio telescope in Ohio, USA. The 'Wow signal' lasted for 37 seconds, and came from part of the sky in the Sagittarius constellation. In the direction in which the radio telescope was pointing, the nearest stars lie 220 light years away. No other interesting signals have been detected from that part of the sky, and the signal remains a mystery to this day.

In 2003, while reviewing the 200 best candidate signals, the SETI@home project looked again at a tiny area of the sky where a potentially interesting signal had been detected twice before. The same pattern of radio emissions was detected for a third time, and it does not resemble anything that radio astronomers have seen before. Moreover, it is at 1420 MHz. However much this result stands out, SETI experts hold out little hope that it will turn out to be the first confirmed extraterrestrial communication.

The idea of SETI has featured in several science-fiction books and films, but none so authentically as the 1985 novel *Contact*, by Carl Sagan. In the story, which was also made into a film in 1997, scientists working on SETI at the Arecibo telescope detect a signal from an alien civilisation. The signal is packed with information on how to build a machine to create a wormhole, so that a representative from Earth can meet with an extraterrestrial. It turns out that a galactic communications network already exists, and Earth has been monitored for some time to see if it is ready to join. While this part of the story might be fantasy, Sagan's study of the reactions of human beings to first contact with extraterrestrials is well informed. Carl Sagan was an astronomer and a consultant to NASA, and he had a keen fascination in the idea of extraterrestrial intelligence.

However authentically *Contact* was put together, the events it contains remain as science fiction for now. The giant radio telescope at Arecibo continues to search daily for alien transmissions, and across the world millions of personal computers are sifting through the data gathered. The search goes on.

Text ET

The Arecibo radio telescope has played a role in the reverse of SETI, sometimes called CETI (Communication with Extraterrestrial Intelligence). In 1974, instead of receiving radio waves from space, the telescope sent some out. The 169-second broadcast was coded as a series of 1,679 binary 'ons' and 'offs'. The number 1,679 can only be divided by two numbers: 23 and 73. It is hoped that if intelligent aliens ever receive the signal, they will arrange the binary digits as a rectangle of white (on) and black (off) squares, with sides of 23 and 73 squares. If they do manage this, they will see a picture that carries information about our planet and our species. The radio signal carrying the message was beamed towards a star cluster 25,000 light years away.

In 1999 the same message was broadcast again, together with some other messages, by a private company called Encounter 2001 (now called Team Encounter). The new, improved message, dubbed the 'Cosmic Call', was directed towards four different stars which had been identified as targets for further research by the SETI Institute, an organisation in California which coordinates the search for extraterrestrial intelligence. Cosmic Call 2 was made, by the same company, in 2003. In 2001 a selection of messages composed by Russian teenagers was beamed to six nearby stars similar to the Sun. Some of the six stars are known to have planets in orbit around them. The first of the interstellar radio messages to reach its destination will do so in the year 2036. If intelligent aliens receive the message, understand it, and send an immediate reply, we will hear from them in 2069.

In addition to intentional radio messages we have transmitted towards distant stars, we have also sent inadvertent ones. Some of our radio, television and radar signals 'leak out' of the atmosphere, travelling away from Earth at the speed of light. They are not very powerful, and weaken dramatically over the large distances in interstellar space. But it is just possible that a very advanced civilisation – somewhere less than 100 light years away – might be able to tease out the signals from the rest of the background radio 'noise' generated naturally by stars, galaxies and interstellar hydrogen gas. This possibility is explored in the film *Contact*, in which the extraterrestrials sending the message also send back television pictures that they have been receiving over many years.

Radio broadcasts are not the only messages we have delivered into space. Metal plaques were attached to both of NASA's Pioneer space probes, which visited several planets before heading into interstellar space. The plaques illustrated information about Earth and human beings – including where to find us. Similarly, both Voyager 1 and 2 carry gold-plated discs that contain speech, music, photographs and diagrams from many different cultures, encoded in grooves like those on a vinyl record. Illustrations on the protective cover of the record explain how to play the record. Sadly, the probability that any intelligent alien will find the Voyager and Pioneer space probes is as close to zero as you could imagine. Space is so vast that the probability of human beings ever meeting face-to-face with other intelligent beings is not much greater. But that does not mean that intelligent aliens are not out there somewhere.

There are at least 100 thousand million (100 billion) stars in our galaxy. Current estimates suggest that planets orbit perhaps half of these stars. The potential variety of living things on these planets is bewildering. And the Galaxy is not the entire Universe: there are at least 100 thousand million galaxies altogether. The total number of stars and planets is huge. Somewhere there must be a planet just like Earth, which had the same atmospheric conditions early on, where extremophiles thrived and led to the development of wonderful plants and animals. Perhaps creatures like dinosaurs evolved, and were wiped out to leave creatures like mammals to dominate. Perhaps one species developed

intelligence like ours, as well as society and culture and technology. Maybe some of those intelligent beings look up into space and wonder whether there is anything else like them.

This vision may be entirely false; it may be that we are the only island of life in the vast ocean of space. But as Jodie Foster's character, Eleanor Arroway, comments in the closing moments of the film *Contact*: 'The Universe is a pretty big place. It's bigger than anything anyone has ever dreamed of before. So if it's just us, it seems like an awful waste of space, right?'

▲ In the film *Contact*, Eleanor Arroway (Jodie Foster) explains to an assembly of politicians and military personnel part of a message received from a distant civilisation.

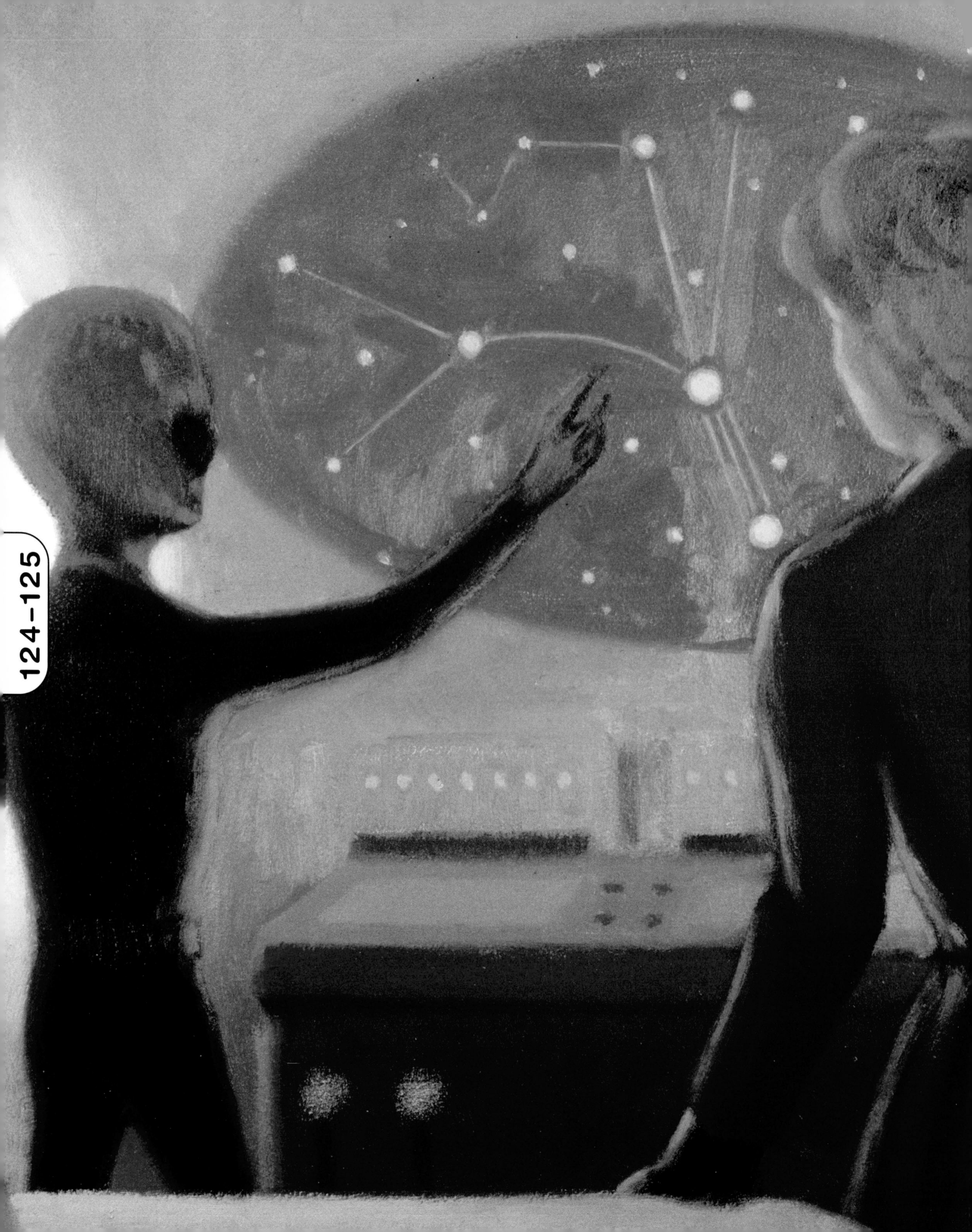

Books

Cohen, Jack and Ian Stewart. *What Does a Martian Look Like?: The Science of Extraterrestrial Life.* London: Ebury Press, 2004.

Davies, Paul. *The Origin of Life.* Harmondsworth: Penguin Books Ltd, 2003.

Fontenelle, Bernard le Bovier de. *Conversations on the Plurality of Worlds.* Trans. H.A. Hargreaves. Berkeley: University of California Press, 1990.

Mayor, Michael and Pierre-Yves Frei. *New Worlds in the Cosmos: The Discovery of Exoplanets.* Trans. Boud Roukema. Cambridge: Cambridge University Press, 2003.

Pickover, Clifford. *The Science of Aliens.* New York: Basic Books, 2000.

Tyson, Neil Degrasse and Donald W. Goldsmith. *Origins: Fourteen Billion Years of Cosmic Evolution.* New York: W.W. Norton and Company, 2004.

Villard, Ray and Lynette R. Cook. *Infinite Worlds: An Illustrated Voyage to Planets Beyond Our Sun.* Berkeley: University of California Press, 2005.

Websites

http://www.daviddarling.info
The most comprehensive and authoritative web-based encyclopedia of astrobiology, astronomy and space flight.

http://www.nasa.gov
The way into NASA's online presence, with a vast collection of images, videos and up-to-date information about past, present and future space missions.

http://www.imdb.com
The Internet Movie Database, 'Earth's biggest movie database'. For all you could ever want to know about science-fiction (and all other) films and television series.

http://www.esa.int
The European Space Agency, with more information about the Darwin project, which aims to find Earth-like planets.

http://www.bis-spaceflight.com
The British Interplanetary Society, a long-running group that promotes and supports space travel.

http://exoplanets.org
A website listing all the known extrasolar planets.

http://www.nineplanets.org
The best web-based source of information about the Solar System.

http://www.seti.org The SETI Institute.

http://seti.ssl.berkeley.edu/serendip/
The homepage of Project SERENDIP, the largest and longest-running search for signals from extraterrestrial civilisations.

http://www.wikipedia.org
The free and extremely authoritative encyclopedia that anyone can edit. Detailed information on any subject you can think of.

http://www.extremophiles.org
The International Society for Extremophiles.

Acknowledgements and Picture Credits

Acknowledgements

The Author and Publishers are grateful to the following for their assistance in the preparation of this book:

Mark Brake is Professor of Science Communication at the University of Glamorgan's Centre for Astronomy and Science Education, and is part of the NASA Astrobiology Science Communication Group. He has been communicating the science of astrobiology and the question of alien life for over 10 years.

Ian Morison is an astronomer at the University of Manchester, Jodrell Bank Observatory. He was co-ordinator of the Project Phoenix SETI observations made in collaboration with the Arecibo Telescope from 1998 to 2003.

Nick Stringer is the Producer/Director of the Big Wave/Channel 4 programme *Alien Worlds*.

Index